Just Transitions

Series Editor
Raphael J. Heffron, University of Pau and Pays de l'Adour, Pau, France

Diane Kraal · Gerry Nagtzaam ·
Susie Siew Yuen Ho · Katie O'Bryan ·
Jadranka Petrovic

A Just Transition to Decarbonisation

Themes of Loss and Damage, Transport, Nature and Youth

palgrave
macmillan

Diane Kraal
Monash Business School
Monash University
Melbourne, Australia

Gerry Nagtzaam
Faculty of Law
Monash University
Melbourne, Australia

Susie Siew Yuen Ho
Office of the Deputy
Vice-Chancellor, Education
Monash University
Melbourne, Australia

Katie O'Bryan
Faculty of Law
Monash University
Melbourne, Australia

Jadranka Petrovic
Monash Business School
Monash University
Melbourne, Australia

ISSN 2731-6041 ISSN 2731-605X (electronic)
Just Transitions
ISBN 978-3-031-49136-8 ISBN 978-3-031-49137-5 (eBook)
https://doi.org/10.1007/978-3-031-49137-5

This Palgrave Macmillan imprint is published by the registered company Springer Nature Switzerland AG
The registered company address is: Gewerbestrasse 11, 6330 Cham, Switzerland

Paper in this product is recyclable.

Preface

This book explores the urgent need for a *just transition* to achieve the goal of decarbonisation. Chapter themes are drawn from the recently held United Nations Climate Change Conferences, or COPs (Conferences of the Parties), that facilitate the *Paris Agreement* negotiations. The selected themes of loss and damage, transport, nature and youth will showcase the knowledge and research expertise of the authors.

The just transition captures the 'justice' process as societies move towards a decarbonised economy. This book delves into climate action through climate justice, energy justice and environmental justice.

Climate justice concerns understanding the burdens of climate change from a human rights perspective; energy justice refers to the application of human rights across the energy lifecycle; and environmental justice aims to treat all citizens equally in the implementation of necessary environmental laws, regulations and policies.

Many of the authors have participated in COPs over the years as university researchers with one of the many types of non-government organizations (NGOs) and environmental non-government organizations (ENGOs). Concurrently with COP government-to-government *Paris Agreement* negotiations, NGO and ENGO meetings are often held with universities, think tanks and other independent groups to engage and contribute to the ongoing climate action debate.

NGOs and ENGOs are among the many constituencies that contribute from the grassroots level to the *Paris Agreement* process. They typically come together at COP public forums to discuss climate issues. These organizations also interact with businesses at country pavilions, in panel presentations in line with special COP themes. For example, in 2022 the International Renewable Energy Association operated a pavilion at COP27 in Egypt for a programme of events that included NGO and sub-national government discussions on an energy-themed project that offers the broader climate community a view of interlinkages between policy and regulation.

In terms of COP themes, in 2021 at COP26 Glasgow, the special theme days included finance, energy, youth; nature, adaptation, loss and damage; gender, science, transport; cities, regions and built environment. In the following year, COP27 Egypt had similar theme days of finance, science, youth and future generations; decarbonisation, adaptation and agriculture; gender, water, *Action for Climate Empowerment* and civil society; energy, biodiversity; and solutions. In late 2023, the COP28 meeting in Dubai, United Arab Emirates, continued these themes that included loss and damage, transport, nature and youth.

COP themes provide an important structure for understanding the need for a just transition to decarbonisation. It is hoped that this book will contribute to understanding this structure and provide its readers with valuable insights. In future, more COP themes can be addressed.

Melbourne, Australia

Diane Kraal
Gerry Nagtzaam
Susie Siew Yuen Ho
Katie O'Bryan
Jadranka Petrovic

Acknowledgments The authors gratefully acknowledge some insights provided by Hailey Campbell; and Nicola Chopin.

Contents

List of Figures

List of Tables

CHAPTER 1

Introduction: A Just Transition to Decarbonisation

Abstract This chapter introduces the journey through this book of a just transition to decarbonisation. It explains the just transition's decarbonisation linkage to United Nations Climate Change Conference themes and provides chapter overviews.

Keywords Just transition · Conference of the Parties conference · Loss and damage · Transport · Nature and youth

This book addresses the challenges of a just transition to decarbonisation through the lens of four United Nations Climate Change Conferences, or Conference of the Parties (COP) themes: loss and damage, transport, nature and youth. These major themes were chosen to provide the reader with a multidimensional understanding of how a just transition is fundamental to all aspects of climate action and intergovernmental decision-making. These themes have become increasingly central to COPs in the last five years. In late 2023, COP28 in Dubai continued these themes, with priorities including launching of the *Just Transition Work*

D. Kraal et al., *A Just Transition to Decarbonisation*, Just Transitions,
https://doi.org/10.1007/978-3-031-49137-5_1

Programme.[1] A just transition has been defined as transitioning away from fossil fuels towards a more just future, to alleviate past social inequalities.[2]

The theme of loss and damage is about funding for delivering the practical solutions needed to adapt to climate impacts. The theme of transport is about driving the global transition to zero-emission transport.

The theme of nature is about ensuring that the importance of natural habitats and sustainable land use is part of global action on climate change and a clean, green recovery. It acknowledges water as the source of life and livelihoods. Climate impacts on water have linkages to wider, cross-cutting impacts on development, livelihoods and Indigenous cultural survival. The theme of nature also has links to international trade—most notably coal-associated trade activity—as it negatively affects climate through greenhouse gas emissions and dramatically exacerbates the condition of the natural environment, including biodiversity. As enforceable free trade agreements (as opposed to the largely voluntary nature of international climate law) are being increasingly negotiated throughout the world, the question of whether these agreements could be used to remedy the tension between the desire for economic development and the need to protect the environment warrants attention.[3]

The theme of youth aims to elevate the voice of young people and demonstrate the critical role of public empowerment and education in climate action. Engagement with youth ensures that their perspectives are taken into account and reflected across all areas of the climate agenda.

The just transition framework seeks to provide the context for policy makers with pathways to deep decarbonisation across global economies to significantly reduce CO_2 emissions by 2030 and beyond. It is acknowledged that there are mixed meanings of a 'just transition' for scholars

[1] Government of Egypt, 'Climate Ministers Meet in Copenhagen for First High-level Meeting Since COP27, Paving the Way for COP28' (Press Release, 20 March 2023). https://reliefweb.int/report/world/climate-ministers-meet-copenhagen-first-high-level-meeting-cop27-paving-way-cop28 ; For the United Nations COP28 themes, see https://climatechampions.unfccc.int/category/cop28/.

[2] D McCauley et al., 'Identifying, Improving, and Investing in National Commitments to Just Transition: Reflections from Latin America and the Caribbean' (2023) 17 *Environmental and Sustainability Indicators* 1.

[3] COP26 Presidency Programme, Thematic Days. https://ukcop26.org/the-conference/presidency-programme/; COP27 Presidency Vision, Thematic Days. https://cop27.eg/#/presidency/eventsThematic.

of climate justice, environmental justice and energy justice.[4] Climate justice is about sharing the benefits and burdens of climate change from a human rights perspective whereas environmental justice aims to treat citizens equally in the development of environmental policies and laws. Energy justice concerns the application of human rights across the energy life cycle.[5] A just transition to achieve decarbonisation covers all three perspectives. The structure and scope of this book's chapters follow next.

Chapter 2, 'Climate Justice and Concept of Loss and Damage in Climate Negotiations' overviews the concept of 'loss and damage', first introduced at the COP19 United Nations climate conference in November 2013 to address losses and damage in developing states. This chapter offers a critical examination of that concept and its potential as a vehicle for developing countries to obtain climate justice.

The increase and severity of destructive climate impacts of anthropogenic climate change has led to questions of attribution; that is, who should pay for historical climate damage? However, ascribing specific losses and damage to human-made climate change has proven a challenge for negotiators. The debate on loss and damage has proven contentious within international climate negotiations.

Chapter 3, 'Zero-Emissions Transport: Taxation support for business electric cars' covers the necessity for taxation support and other fiscal subsidies necessary to increase the number of zero-emission vehicles in business fleets in Australia. Road transport contributed to 19% of Australia's CO_2 emissions in 2022, with passenger vehicles contributing almost 50% of those emissions. In Australia, 40% of light passenger vehicles sold in 2020 were purchased by businesses. Almost half of business fleet vehicles are home garaged, providing an opportunity for fleet electric vehicles to be home charged using smart meters to gain off-peak rates, avoid grid congestion and address the lack of business-site charging. This chapter also provides European tax comparisons and then explains and analyses legal options for achieving reduced emissions from business

[4] RJ Heffron, *Achieving a Just Transition to a Low-Carbon Economy* (Springer International Publishing AG, 2021).

[5] D Kraal, V Haritos and R Cantley-Smith, 'Tax Law, Policy and Energy Justice: Rethinking Biofuels Investment and Research in Australia' (2020) 35(1) *Australian Tax Forum* 31; D Kraal, 'Petroleum Industry Tax Incentives and Energy Policy Implications: A Comparison between Australia, Malaysia, Indonesia and Papua New Guinea' (2019) 126 (March) *Energy Policy* 212.

fleets through consideration of Australian tax incentives for zero-emission vehicles.

Chapter 4, 'Water is Life—Recognising First Nations in Sustainable Water Management and Use' focuses on the COP theme of nature, and highlights the importance of sustainable water use and management in tackling climate change, and the vital role of First Nations people in that space.

First Nations people have accumulated, via their ancestors, many thousands of years of traditional knowledge about water use and management. However, the value of this knowledge is often given only lip service in Western water management regimes. Further, the effects of climate change disproportionately impact Indigenous communities because of their spiritual and cultural relationship with the environment and their dependence on land and water for their livelihoods. Accordingly, this chapter examines whether the regulatory framework for water use and management in Australia adequately provides for Indigenous participation in decision-making.

Chapter 5, 'Utilising Free Trade Agreements to Protect Nature and Heritage' is concerned with the role of free trade agreements in redressing the harm inflicted upon the natural environment, and in contributing to a just transition to a low emission and resilient economy. While essential for nations' economic development, the continuous vast expansion of international trade has significantly impacted the environment, including the ecosystems of areas purportedly designated to be preserved for the benefit of all humankind. As free trade agreements are being increasingly negotiated throughout the world, the question of their utility vis-a-vis the environment is becoming more important. By taking the world heritage approach, this chapter argues that free trade agreements have a crucial role to play in remedying the harm wrought on the environment and in ensuring that the biodiversity we have inherited is transmitted to future generations.

Chapter 6, 'Youth Empowerment for a Just Transition', considers the need for the voice of youth to be louder than ever before. Youth are frustrated and want action but do not always have the skills and knowledge to achieve it. At COPs, many non-governmental stakeholders have discussed their concerns about whether a diversity of voices from the public are being appropriately represented in climate-related decision-making. As our physical, social and economic world continues to change from historical baselines, the most vulnerable among us—such as youth, women, children and Indigenous communities—will be most impacted.

The climate crisis is not experienced equitably around the globe and youth voices consistently rate this as a top concern. Youth, First Nations peoples, people of colour, persons with disability, women and LGBTIQ + communities bear the brunt of extreme climate events, yet these same voices remain underrepresented among COP scientists, leaders and decision-makers. The COP is a unique pathway to develop diverse future leaders in global governance, intercultural liaison, intergovernmental policy and career-defining collaborations. This chapter argues for broader access to educational tools and tailored curricula that build negotiation skills and networked opportunities at global fora.

Chapter 7, the 'Conclusion', discusses the ways in which the chapters fit with the focus of the book; that is, to advance the need for a just transition to decarbonisation. It overviews how each chapter relates to the selected themes from the COPs that facilitate *Paris Agreement* negotiations. The chapter summarises the way forward for each COP theme.

Some chapters have a specific focus on Australia, while others draw on international literature. Nonetheless, all the authors acknowledge the global effort required for a just transition to decarbonisation.

CHAPTER 2

Climate Justice and Concept of Loss and Damage in Climate Negotiations

Abstract To achieve a just transition to decarbonisation, considerable amounts of investment and financial resources are needed. Therefore, the agreement to a loss and damage fund at the United Nations Climate Change Conference in Egypt (COP27) is an important moment in the global climate justice movement. Climate justice is a normative framework that seeks to combat fundamental inequities arising from the ongoing climate crisis, and may encompass concepts such as distributive justice, recognitional justice and representational justice. The COP27 initiative of an L&D fund has been said to formally address the fundamental injustice underlying climate negotiations: that the countries least responsible bear the heaviest burdens from climate change and have not, until now, received adequate assistance from those most responsible. This chapter recounts the history of the agreement to establish a loss and damage fund and its framework.

Keywords Loss and damage · Climate justice · Climate negotiations · COP27

D. Kraal et al., *A Just Transition to Decarbonisation*, Just Transitions, https://doi.org/10.1007/978-3-031-49137-5_2

2.1 Introduction

Historically there has been recognition in the international climate community that, while the effects of anthropogenic climate change are global, some states are particularly at risk. These vulnerable states[1] have populations that experience high rates of poverty and marginalisation and/or are geographically located in areas that are particularly susceptible to extreme weather events.[2] Further, they are less economically capable when it comes to coping with climate-related crises, which hinders their recovery from disasters and worsens the impacts of damage.[3]

For Lesser Developed Countries (LDCs) and Small Island Development States (SIDS), the situation is further exacerbated in that the physical impacts of anthropogenic climate change are felt unevenly across nations. For example, countries such as Bangladesh have very little responsibility for global greenhouse gas (GHG) emissions and because they have relatively weak economies, will suffer greater impacts. SIDS, however, are responsible for less than 1% of global GHG emissions,[4] and LDCs, for only 4%.[5] Conversely, developed countries such as the United

[1] As a result of increasing anthropogenic climate change, the poor communities in Lesser Developed Countries and the Small Island Developing States are considered more vulnerable to extreme weather events and sea level rising with their concomitant physical and human health costs; AA Faruque, *Environmental Law: Global and Bangladesh Context* (New Warsi Book Corporation, 2020) 110. The Intergovernmental Panel on Climate Change's (IPCC's) Sixth Assessment Report states that Small Island Developing States are particularly vulnerable to climate change because of their geographic locations and high levels of exposure to climate impacts, as well as low levels of economic diversification, governance arrangements, technology development, financial resources and capacity limitations in institutional and legal systems; IPCC 2021, 'Summary for Policymakers', in V Masson-Delmotte et al., *Climate Change 2021: The Physical Science Basis* (Contribution of Working Group I to the Sixth Assessment Report of the IPCC, Cambridge University Press, 2021).

[2] Patrick Toussaint, 'Voices Unheard—Affected Communities and the Climate Negotiations on Loss and Damage' (2018) 3(5–6) *Third World Thematics* 768.

[3] Birsha Ohdedar, 'Loss and Damage from the Impacts of Climate Change: A Framework for Implementation' (2016) 85(1) *Nordic Journal of International Law* 1, 8.

[4] United Nations Development Programme, *Small Island Developing States: The State of Climate Ambition* (Report, December 2022) 3.

[5] United Nations Conference of Trade and Development, *The Least Developed Countries Report 2022* (Report, 2022) 8.

States (US) contribute more GHGs and face far less immediate threats of impact.[6]

Thus, it has become increasingly clear to the global community that there is a 'vast disjuncture' between those who are responsible for human-induced climate change and those who will be, and in some cases are already, suffering the greatest consequences.[7] Asking these vulnerable states to bear their own costs from a crisis that others have caused has been labelled the 'ultimate manifestation of climate injustice'.[8]

To date, efforts by state parties to the *United Nations Framework Convention on Climate Change* ('*UNFCCC*') and subsequent protocols (e.g. Kyoto) and agreements (e.g. Paris) to combat the onset and effects of anthropogenic climate change have predominantly involved state-level mitigation[9] and adaptation[10] measures being sought to be implemented by nations.[11] However, with the growing fear that mitigation and adaptation strategies are not sufficient in the case of vulnerable states,[12] there has been a growing recognition of the necessity for a separate mechanism to address climate consequences, termed climate-related loss and damage

[6] Patrick Toussaint, 'Loss and Damage and Climate Litigation: The Case for Greater Interlinkage' (2021) 30 *Review of European, Comparative and International Environmental Law* 16.

[7] Ohdedar (Footnote 3) 7.

[8] Ferreira Galvão, 'Arrested Development: The Late and Inequitable Integration of Loss and Damage Finance into the UNFCCC' in Meinhard Doelle (ed), *Research Handbook on Climate Change Law and Loss & Damage* (Edward Elgar Publishing, 2021) 127, 127.

[9] Mitigating climate change strategies are designed to reduce the release of greenhouse gases, which are warming the planet. Such strategies include retrofitting buildings to make them more energy efficient and adopting renewable energy sources; for example, solar and wind. https://www.thegef.org/what-we-do/topics/climate-change-mitigation.

[10] Adaptation strategies are about society and the planet adapting to life in a changing climate. They involve adjusting to actual or expected future climate impacts with the goal of reducing the inherent harmful effects of anthropogenic climate change; for example, sea level rise, increased rates of extreme weather events and global food insecurity. https://climate.nasa.gov/solutions/adaptation-mitigation/.

[11] Jorge Gabriel Arevalo Garcia, 'Challenges of Compensation and Reparation for Loss and Damage Related to the Adverse Effects of Climate Change' (2020) 13(1) *Mexican Law Review* 190; Edward A Page and Clare Heyward, 'Compensating for Climate Change Loss and Damage' (2017) 65(2) *Political Studies* 356.

[12] A 2022 report by the IPCC argues that when the recognisable limits of mitigation and adaptation are exceeded, loss and damage (L&D) occurs. IPCC, *Climate Change 2022: Mitigation of Climate Change*, IPCC AR6 WGIII (4 April 2022) 2430.

(L&D).[13] Though historically not formally defined in any legal text, L&D can initially be understood to encapsulate the negative effects of climate change—both economic and non-economic—that occur despite mitigation and adaptation measures (see the definition section for an in-depth discussion of the term).[14]

LDCs argue that developed, high-emitting countries should be held responsible for the L&D currently being caused and that which will occur in the future. To that end, vulnerable states have fought to create a 'loss and damage finance facility'. An L&D fund would be financed by wealthier, industrialised 'developed' countries who are regarded as primarily responsible for anthropogenic climate change through their historical output of carbon emissions.[15] This line of argument stresses that because of their actions, developed states have a moral and legal obligation to compensate vulnerable countries.[16] However, these claims for compensation have evoked a defensive reaction by developed countries, as such claims place liability on them for climate change-derived L&D—which many deny—as well as potentially legally requiring them to provide just compensation for past actions.[17]

It has been estimated that developing states could experience L&D worth up to US$400 billion annually by 2030 even if global warming is curtailed below 2 °C.[18] By focussing on both financial and non-financial dimensions, the proposed L&D fund seeks to offer substantial financial support to vulnerable nations (LDCs and SIDS), recognising the

[13] Garcia (Footnote 11) 190.

[14] Liselotte Jensen and Paulina Jabczyńska, *Understanding Loss and Damage: Addressing the Unavoidable Impacts of Climate Change* (Briefing Report, European Parliamentary Research Service, 13 July 2022) 2.

[15] Gregor Vulturius and Marion Davis, *Defining Loss and Damage: The Science and Politics Around One of the Most Contested Issues Within the UNFCCC* (Stockholm Environment Institute, 2016). https://mediamanager.sei.org/documents/Publications/Climate/SEI-DB-2016-Loss-and-damage-4-traits.pdf.

[16] Toussaint (Footnote 6).

[17] World Resources Institute, 'COP27's Success Hinges on Loss and Damage Finance Discussions', *United Nations Development Programme* (Web Page, 28 October 2022). https://www.undp.org/blog/what-new-loss-and-damage-fund-needs-success.

[18] Jonathan Gewirtzman et al., 'Financing Loss and Damage: Reviewing Options Under the Warsaw International Mechanism' (2018) 18(8) *Climate Policy* 1076, 1078.

critical nature of the issues faced by such states in addressing the multi-faceted challenges of anthropogenic climate change in an equitable and just manner.[19]

To achieve a just transition to decarbonisation, considerable amounts of investment and financial resources are needed. Thus, the creation of an L&D fund at the United Nations Climate Change Conference, or Conference of the Parties (COP) in Egypt in 2022 (COP27) has been labelled an 'important moment in the global climate justice movement'.[20] The issue of 'climate justice' has been a consistent topic of debate during international climate negotiations despite ongoing debate around both defining and implementing the concept.[21] At its simplest, climate justice refers to the reality that climate change most severely impacts those with the least responsibility for causing it.[22] It therefore frames anthropogenic climate change as a moral and social justice issue, rather than just a scientific, technical or financial problem to be solved.[23] A central principle of the concept is that, 'those who are suffering the most from the impacts of climate change, although they have contributed the least to global warming, deserve financial support from those who have ... done the most damage to the environment'.[24]

Climate justice is therefore a normative framework that seeks to combat fundamental inequities arising from the ongoing climate crisis, and may encompass concepts such as distributive justice, recognition

[19] Elisabeth Mahase, 'Climate Change: "Loss and Damage" Fund Payouts Could Take Decades, Scientists Warn' (2022) 379 *British Medical Journal* 1.

[20] Arthur Wyns, 'COP27 Establishes Loss and Damage Fund to Respond to Human Cost of Climate Change' (2022) 7(1) *The Lancet Planetary Health* 1. https://www.thelancet.com/journals/lanplh/article/PIIS2542-5196(22)00331-X/fulltext.

[21] Chukwumerije Okereke and Philip Coventry, 'Climate Justice and the International Regime: Before, During, and After Paris' (2016) 7(6) *WIREs Climate Change* 834, 835.

[22] Peter Newell, 'Climate Justice' (2022) 49(5) *The Journal of Peasant Studies* 915, 916; Ohdedar (Footnote 3) 6.

[23] Farhana Sultana, 'Critical Climate Justice' (2022) 188(1) *The Geographical Journal* 118; Stockholm Environment Institute, *Operationalizing Finance for Loss and Damage: From Principles to Modalities* (Report, November 2022) 11.

[24] Bilawal Bhutto Zardari, 'The Loss and Damage Facility: A Step Towards Climate Justice' *UN Chronicle* (Blog Post, 9 December 2022). https://www.un.org/en/un-chronicle/loss-and-damage-facility-step-towards-climate-justice#:~:text=generations%20yet%20unborn.-,While%20the%20agreement%20does%20not%20establish%20the%20legal%20responsibility%20of,have%20contributed%20the%20least%20to.

justice and representational justice.[25] An L&D fund has been said to formally address the fundamental injustice underlying climate negotiations: that the countries least responsible bear the heaviest burdens from climate change and have not, until now, received adequate assistance from those most responsible.[26] By focussing on both financial and non-financial dimensions, the mooted L&D fund seeks to provide substantial support to vulnerable nations while recognising the imperative nature of addressing the 'wicked problem' of climate change in an equitable and just manner.[27] Thus, an L&D fund is potentially a huge step in implementing climate justice.[28]

The position of L&D has experienced long-standing definitional and operational uncertainty. This chapter explores the differing definitions of L&D extent and proposes the adoption of a particular version of the term that should be acceptance by the international community. It then traces the term's historical usage in United Nations (UN) negotiations leading to the current landscape of the topic today. The chapter then analyses in detail the negotiation and key takeaways relating to L&D from COP27 in Egypt in 2022, which are framing the L&D agenda for 2023's COP28.[29] For the first time, member-states agreed to operationalise a climate L&D fund, specifically focussing on vulnerable

[25] Anastassia Claire Tikhomirova, 'Establishing a Loss and Damage Fund: How Small Island Developing States Negotiated a "Historic Deal for Climate Justice"' (Bachelor Thesis, Malmo University, 2023) 11, 25. In this context, distributive justice can be defined as the equitable division of the economic 'costs and burdens' associated with climate change; recognition justice promotes the recognition of the social, cultural and historical contexts, such as colonisation, which may lead to increased levels of vulnerability; and finally, representational justice advocates for the increased involvement of relevant stakeholders in decision making regarding loss and damage, including vulnerable countries, indigenous communities and other members of the private sector.

[26] Ibid.

[27] Mahase (Footnote 19) 1.

[28] Tal Harris, 'COP27 Loss and Damage Finance Fund a Down Payment on Climate Justice', *Greenpeace* (Blog Post, 22 November 2022). https://www.greenpeace.org/africa/en/press/52751/cop27-loss-and-damage-finance-fund-a-down-payment-on-climate-justice; Bilawal Bhutto Zardari, 'The Loss and Damage Facility: A Step Towards Climate Justice', *UN* (Blog Post, 9 December 2022). https://www.un.org/en/un-chronicle/loss-and-damage-facility-step-towards-climate-justice.

[29] 'COP27 Ends with Announcement of Historic Loss and Damage Fund', *United Nations Environment Programme* (Web Page, 22 November 2022). https://www.unep.org/news-and-stories/story/cop27-ends-announcement-historic-loss-and-damage-fund.

nations.[30] However, as a number of key questions about the fund's scope have yet to be answered, the chapter concludes with a critical examination of the outstanding issues confronting climate negotiators regarding L&D if it is to become an effective vehicle for climate justice.

2.2 Defining the Concept of Loss and Damage

There is no universally agreed definition for the term L&D, primarily because of competing perspectives on the concept.[31] This normative struggle has led to stakeholders addressing the concept differently depending on their perspective or objective in seeking to have their interpretation adopted as the preferred nomenclature.[32] Problematically, and adding to the normative confusion around the phrase, no official definition of L&D was adopted under the *UNFCCC*.[33] Further complicating the issue is that the term has proven vague and ill-defined as framed in climate policy discussion.[34] For some, such 'constructive ambiguity' is critical for allowing negotiations to continue unabated as this vagueness can be elucidated in the future.[35]

Several competing definitions have been advanced by stakeholders and academics. In 2012, a *UNFCCC* literature review defined L&D as, 'the actual and/or potential manifestation of impacts associated with climate change...that negatively affect human and natural systems'.[36] However, that definition has not been widely adopted by states and key stakeholders,

30 Ibid.

31 Toussaint (Footnote 6) 18.

32 Richard Tol and Roda Verheyen, 'State Responsibility and Compensation for Climate Change Damages a Legal and Economic Assessment' (2004) 33 *Energy Policy* 13.

33 *United Nations Framework Convention on Climate Change* opened for signature 9 May 1992, 1771 UNTS 107 (entered into force 21 March 1994) art 3(1) *('UNFCCC')*; Jessica Baker, 'Australia Is Burning: Aligning Corporate Social Responsibility and Community Expectations Following the Black Summer' (2021), 36 *Australian Journal of Corporate Law* 113.

34 Jensen and Jabczyńska (Footnote 14) 1; Lisa Vanhala and Cecilie Hestbaek, 'Framing Climate Change Loss and Damage in UNFCCC Negotiations' (2016) 16(4) *Global Environmental Politics* 111, 112.

35 Itay Fischhendler, 'When Ambiguity in Treaty Design Becomes Destructive: A Study of Transboundary Water' (2008) 8(1) *Global Environmental Politics* 111.

36 UNFCCC, *A Literature Review on the Topics in the Context of Thematic Area 2 of the Work Programme on Loss and Damage: A Range of Approaches to Address Loss and*

and remains contested.[37] Thus, multiple definitions have been suggested by those seeking to both widen and delimit the ambit of any discussions on the issue. A broad interpretation potentially includes 'all negative consequences of climate change', whereas a narrow definition is limited to the 'impacts of climate change that have not been, or cannot be, avoided through mitigation or adaptation efforts'.[38]

Problematically, each proffered definition has its own scope of what should be considered L&D, leading inevitably to both divergent priorities and differing understandings of the urgency with which the term 'L&D' should be addressed.[39] Van der Geet and Warner stress that L&D has no clear definition, but can be understood as the consequences of anthropogenic climate change that can no longer be mitigated against or are beyond adaptation strategies.[40] Similarly, Vulturius and Davis argue that any working definition of L&D incorporates the adverse residual impacts of climate change that go beyond the acceptable limits of mitigation and adaptation efforts and can be best understood as unavoidable, irreversible, attributable and intolerable.[41] Toussaint expands the definition to argue that, 'the consequences of climate change that go beyond what people can adapt to, or when options exist but a community does not have the resources to access them'.[42] Such economic, social and cultural losses have historically been disproportionately experienced by LDCs and SIDS.[43]

Damage Associated with the Adverse Effects of Climate Change, 37th session, FCCC/SBI/2012/INF.14 (15 November 2012), 4[7].

[37] For example, Jensen and Jabczyńska argue the definition of L&D should be, 'impacts of climate change that have not been or cannot be avoided through mitigation or adaptation efforts' (Footnote 14) 2.

[38] Ibid.

[39] Morten Broberg and Beatriz Martinez Romera, 'Loss and Damage After Paris: More Bark Than Bite?' (2020) 20(6) *Climate Policy* 662.

[40] Kees van der Geet and Koko Warner, 'Loss and Damage from Climate Change: Emerging Perspectives' (2015) 8(2) *International Journal of Global Warming* 133, 135.

[41] Vulturius and Davis (Footnote 15).

[42] Toussaint (Footnote 6) 17; Preety Bhandari et al., 'What Is "Loss and Damage" from Climate Change? 6 Key Question, Answered', *World Resources Institute* (Web Page, 6 April 2022). https://www.wri.org/insights/loss-damage-climate-change.

[43] Sarahi Nunez Ramos and Matteo De Felice, 'COP27: The Past, the Present and the Future of Loss and Damage Compensation' *Rabobank*. https://www.rabobank.com/knowledge/d0113373310-cop27-the-past-the-present-and-the-future-of-loss-and-damage-compensation.

At its core, the term can thus be understood as referring to the 'residual effects' of anthropogenic climate change that cannot be forestalled by mitigation or adaptation strategies.[44] Broberg and Martinez put forth a more subjective definition seeing L&D as stemming primarily from risks that cannot be kept within a determined range of 'tolerable risks'.[45] L&D unambiguously arises when anthropogenic climate change impacts exceed tolerable adaptation limits.[46] The express exclusion of adaptation from current working definitions is no doubt a result of its ex post evaluation of harm. Efficacious adaptation strategies would create the resilience necessary to prevent the projected adverse climate impacts.[47]

For Doelle and Seck, the concept denotes two categories of harm associated with anthropogenic climate change impacts. First, any 'loss' incurred references permanent or irrecoverable harm, such as the loss of landmass from sea level rise. Second, the concept necessitates the inclusion of non-economic considerations; for example, traditional knowledge, human mobility and social disruption.[48] While economic losses are tangible and more easily quantifiable, non-economic losses can be more devastating and require financial assistance.[49]

2.2.1 Framing the Debate on the Meaning of Loss and Damage

To serve their competing state interests, both developed countries and LDCs have adopted competing frameworks through which to interpret the goals, boundaries and nature of the term, and involving questions of value judgements and interpretations of justice.[50] Not surprisingly given the stakes involved, LDCs have sought to advance the broadest definition, arguing L&D should be considered as all impacts of climate change

[44] Tol and Verheyen (Footnote 32) 13.

[45] Broberg and Romera (Footnote 39) 662.

[46] Kirsten Dow et al., 'Limits to Adaptation' (2013) 3(4) *Nature Climate Change* 305.

[47] IPCC, 'Summary for Policymakers' (Footnote 1) 3.

[48] Meinhard Doelle and Sara Seck, 'Loss & Damage from Climate Change: From Concept to Remedy?' (2020) 20(6) *Climate Policy* 669. https://www.tandfonline.com/doi/full/10.1080/14693062.2019.1630353.

[49] Bhandari et al. (Footnote 42).

[50] Toussaint (Footnote 6).

residual to mitigation and adaptation.[51] For LDCs, L&D arguments are advanced within a liability and compensation frame under which developed states they maintain to have primarily contributed to anthropogenic climate change through the large-scale emission of GHGs are considered accountable for the resulting harms and therefore have an obligation to compensate victims.[52]

Developed states have vociferously opposed the concept being adopted within climate negotiations. These nations fear being forced to accept responsibility and liability for compensation to be repaid to LDCs that have been impacted because of industrialised countries' historical carbon emissions.[53] Consequently they have continually sought to avoid any assignment of liability for their prior actions. If forced to consider such an approach, they argue it should be limited to a risk management and resilience-building framework. Any financial assistance rendered should be voluntary rather than a compulsory charge levied under a legal obligation.[54] See, for example, US Climate Envoy John Kerry's statement in November 2021 that the US will not allow a legal track to develop within climate negotiations with respect to any liability regarding L&D.[55] Similarly, the Australian government's submission to the first global stocktake in April 2022 only committed to 'considering' how L&D could be addressed by the state at an unspecified future time.[56]

UNFCCC parties need to address the ambiguities surrounding the definition of L&D. They should do this by adopting an official, clear and precise definition, as well as giving L&D a standing place on the

[51] Broberg and Romera (Footnote 39) 662.

[52] Jensen and Jabczyńska (Footnote 14) 4. For example, Vanuatu declared that states suffering L&D are entitled to be compensated. If United Nations processes do not provide such relief, affected states are within their rights to pursue legal action against major emitters. Government of Vanuatu, 'Vanuatu's Revised and Enhanced 1st Nationally Determined Contribution 2021–2030' *Submission to the United Nations Framework Convention on Climate Change* (9 August 2022) 35.

[53] Vanhala and Hestbaek (Footnote 34) 111.

[54] Jensen and Jabczyńska (Footnote 14) 4.

[55] The White House, 'On-The-Record Press Call by Special Envoy John Kerry and National Climate Advisor Gina McCarthy Previewing Day One of COP26' (Press Briefing, 1 November 2021).

[56] Government of Australia, *Submission to the United Nations Framework Convention on Climate Change, First Technical Dialogue of the Global Stocktake* (1 April 2022) 1, 4.

COP agenda.[57] Doing so would reduce the number of current diverging perspectives, a key step in creating the conditions necessary for unified action on the issue. It is submitted here that parties should adopt the 'beyond adaptation' approach, which defines L&D as actions designed to address all adverse effects of anthropogenic climate change that occur despite ongoing mitigation and adaptation measures.[58]

A key advantage of such an approach is that it is broad enough to allow for a comprehensive response to the complex and interdependent range of climate change effects. Impacts left unaddressed by current global mitigation and adaptation strategies would thus be recognised and remedied through adopted-as-needed L&D mechanisms. Such a definition can create a distinct boundary between adaptation and L&D approaches, allowing for less subjectivity than under alternative proposed formulations such as the 'tolerable risk' approach promoted by some advocates.[59]

2.3 History of the Concept in United Nations Climate Negotiations

Because of their geographical location, SIDS have 'disproportionate and unique vulnerabilities to climate change'.[60] Consequently they have chosen to group together in climate negotiations, seeking a more influential role in proceedings with one focal issue being compensation for climate damage.[61] Proposals for a standalone L&D mechanism within the *UNFCCC* process date back to the drafting process in 1991 when the Alliance of Small Island States (AOSIS), represented by Vanuatu, called for the establishment of a fund to compensate vulnerable nations affected

[57] In an open letter to the heads of negotiating delegations, the Climate Action Network called for 'L&D', or at least 'Finance for L&D', to become a standing item on the COP agenda alongside mitigation and adaptation. Climate Action Network, 'Letter to the Heads of Delegation: Agree on the Loss and Damage Finance Agenda Item for the COP27 Climate Conference' (Advocacy Letter, 18 August 2022) [8].

[58] Reinhard Mechler et al., *Loss and Damage from Climate Change: Concepts, Methods and Policy Options* (Springer International Publishing, 2019) 41.

[59] Broberg and Romera (Footnote 39) 662.

[60] Athaulla Rasheed, 'Role of Small Islands in UN Climate Negotiations: A Constructivist Viewpoint' (2019) 56(4) *International Studies* 215.

[61] Ibid. 227.

by anthropogenic climate change.[62] Vanuatu proposed that such a fund should be in the form of an insurance scheme comprised of mandatory contributions from wealthy states to compensate SIDS for adverse impacts resulting from rising sea levels, arguing that the[63]:

> financial burden of loss and damage suffered by the most vulnerable small island and low-lying developing countries … as a result of sea level rise shall be distributed in an equitable manner amongst the industrialised developed countries.[64]

The proposal, however, was poorly received by developed states for several reasons. First, the wealthy states were wary of incurring such potentially large financial obligations. Second, they did not recognise the term as a current issue; rather perceiving it as a problem to be dealt with in the future. Last, the focus of the *UNFCCC* negotiations at that time was creating a binding international agreement to reduce GHG emissions.[65] Thus, for over a decade the focus of the negotiations was mitigation (and later adaptation); L&D was given short shrift.[66]

62 Intergovernmental Negotiating Committee (INC) [Vanuatu: Draft Annex Relating to Article 23 (Insurance) for the Inclusion in the Revised Single Text on Elements Relating to Mechanisms (A/AC.237/WG.II/Misc.13)] 17 December 1991; Broberg and Romera (Footnote 39) 661. Since its creation at the Second World Climate Conference, the Alliance of Small Island States (AOSIS) has advocated for both aggressive and transformational climate mitigation efforts from all states. Erin Roberts and Mark Pelling, 'Loss and Damage: An Opportunity for Transformation?' (2020) 20(6) *Climate Policy* 758, 759.

63 INC [Vanuatu: Draft Annex Relating to Article 23 (Insurance) for Inclusion in the Revised Single Text on Elements Relating to Mechanisms, Intergovernmental Negotiating Committee for a Framework Convention on Climate Change, 4th session, Agenda Item 2(b), A/AC.237/WG.II/CRP.8 (17 December 1991).

64 Insurance Mechanism, Annex, para. 1(1), UN Doc. A/AC.237/WG.II/CRP.8 (17 December 1991) [3(1)]. https://unfccc.int/sites/default/files/resource/docs/a/wg2crp08.pdf.

65 John Ashe, Robert Lierop and Anilla Cherian, 'The Role of the Alliance of Small Island States (AOSIS) in the Negotiation of the United Nations Framework Convention on Climate Change (UNFCCC)' (2009) 23(3) *Natural Resources Forum* 209; Cynthia Liao et al., 'What Is Loss and Damage?', *Chatham House* (Web Page, 6 December 2022). https://www.chathamhouse.org/2022/08/what-loss-and-damage#:~:text=Loss%20and%20damage%20was%20first,impacted%20by%20rising%20sea%20levels.

66 Julia Kreienkamp and Lisa Vanhala, 'Climate Change Loss and Damage' (Policy Brief, Global Governance Institute, March 2017) 5. https://www.ucl.ac.uk/global-governance/sites/global-governance/files/policy-brief-loss-and-damage.pdf.

The idea of L&D was reintroduced and formally recognised at the 2007 Bali COP (COP13) as part of the *Bali Action Plan*, according to which state parties agreed to develop, 'means to address loss and damage associated with climate change impacts in developing countries'.[67] Vulnerable states had adopted a unified position demanding the inclusion of L&D within the regime as they were increasingly being impacted by climate-induced disasters.[68] However, despite the inclusion of L&D in the *Bali Action Plan*, developed states at COP14 made it clear that they remained adamantly opposed to any policy of compensation and rehabilitation.[69]

The formal recognition of L&D in 2007 emboldened LDCs to seek to entrench and develop the idea. Between 2008 and 2010, successive proposals on insurance, risk management and compensation were put forth by AOSIS at the climate negotiations.[70] At COP16 in Cancun, Mexico, L&D was incorporated into a two-year work programme under the *Cancun Adaptation Framework*. However, L&D at that time remained subsumed under adaptive policy. The framework allowed for the assessing of L&D risks; investigating appropriate strategies for addressing L&D; and clarifying the role of the *UNFCCC* process in implementing such mechanisms. Emboldened by these successes, vulnerable states became increasingly active in building momentum for L&D as a separate concept.[71]

2.3.1 Warsaw International Mechanism

In 2013, at COP19 in Warsaw, Poland, the failure to advance L&D talks saw intense and volatile debate as LDCs were frustrated with the weak progress to date on the issue, leading to 132 states arguing they were

[67] Bali Action Plan, 2007 [1(c)(iii)]; Liao et al. (Footnote 65). https://www.unisdr.org/files/5526_BaliActionPlanandDRR.pdf.

[68] Kreienkamp and Vanhala (Footnote 66); Karen McNamara and Guy Jackson, 'Loss and Damage: A Review of the Literature and Directions for Future Research' (2019) 10(2) *Wiley Interdisciplinary Review* 564.

[69] Kreienkamp and Vanhala (Footnote 66) 11.

[70] AOSIS, 'Proposal to the AWG-LCA: Multi-Window Mechanism to Address Loss and Damage from Climate Change Impacts' (2008); Tuvalu, 'Draft "Copenhagen" Protocol to the United Nations Framework Convention on Climate Change' (2009).

[71] Vanhala and Hestbaek (Footnote 34) 111.

being ignored on the topic and exiting the conference.[72] Parties to the *UNFCCC* subsequently agreed to establish 'institutional arrangements such as an international mechanism' on L&D, leading to the establishment of the *Warsaw International Mechanism for Loss and Damage associated with Climate Change Impacts* ('*WIM*').[73] The overarching aim of the *WIM* is to enable developing states to deal with L&D experienced because of the adverse effects of anthropogenic climate change.[74] The *WIM* was established to fulfil the role of the *UNFCCC* and given three essential functions: (1) increase knowledge and expertise; (2) strengthen ongoing dialogue and coordination; and (3) enhance action and support to address L&D.[75] Critically, enabling the *WIM* established a space within climate negotiations for L&D policy to be further refined and entrenched. It also ensured that L&D discussions would encapsulate both extreme weather events and slow-onset events.[76] Significantly, however, the *WIM* also recognises that climate L&D cannot always be reduced by adaptation alone.[77]

The *WIM* was only 'greenlit' when LDCs dropped previous demands for compensation, because of ongoing concerns among wealthier states that liability for anthropogenic climate change would have costly repercussions for international law.[78] Developed states agreed to the *WIM* going into operation only if it remained under the auspices of adaptation and the Cancun Framework, rather than being designated a third, separate

[72] Tom Prater and Josh Gabbatiss, 'Timeline: The Struggle Over "Loss and Damage" in UN Climate Talks', *Carbon Brief* (Online, 27 September 2022). https://interactive.carbonbrief.org/timeline-the-struggle-over-loss-and-damage-in-un-climate-talks/.

[73] UNFCCC, *Report of the Conference of the Parties on its Eighteenth Session, Held in Doha From 26 November to 8 December 2012: Part Two: Action Taken by the Conference of Parties at its Eighteenth Session*, FCCC/CP/2012/8/Add.1, Decision 3/CP.18.

[74] UNFCCC, 2013 Decision 2/CP.19: Warsaw International Mechanism for Loss and Damage associated with Climate Change Impacts FCCC/CP/2013/10/Add.1.

[75] UNFCCC, 'Warsaw International Mechanism for Loss and Damage associated with Climate Change Impacts (WIM)', *United Nations Climate Change* (Web Page). https://unfccc.int/topics/adaptation-and-resilience/workstreams/loss-and-damage/warsaw-international-mechanism.

[76] Liao et al. (Footnote 65).

[77] Preamble, *United Nations Framework Convention on Climate Change*, Decision 2/CP.19, UN Doc FCCC/CP/2013/10/Add.1 (23 November 2013).

[78] Veera Pekkarinen, Patrick Toussaint and Haro van Asselt, 'Loss and Damage After Paris: Moving Beyond Rhetoric' (2019) 13(1) *Carbon & Climate Law Review* 31, 36–37.

pillar of the *UNFCCC*.[79] LDC negotiators have defended the creation of the *WIM* by arguing that failure to segregate L&D from adaptation is an issue of mere normative value and has no practical value.[80]

Unfortunately, after a decade of the *WIM* in operation as an interim measure it is clear it has achieved little of substance to enhance action to address the ongoing L&D experienced by vulnerable states.[81] Its aspirational transformative potential has been undermined by the limited provision of resources by developed nations.[82] The *WIM* has failed to hold industrialised economies accountable for historical and ongoing anthropogenic climate change. It has also failed to achieve tangible commitments from wealthier states to help LDCs cope with L&D, beyond sharing knowledge without concomitant sharing of capacities.[83] The developed nations were adamant that the financing target of US$100 billion negotiated as per the 2009 *Copenhagen Accord* to cover mitigation and adaptation activities would not be reopened to allow for L&D costs.[84]

2.3.2 Paris Agreement 2015

Leading up to COP21 in Paris, France, in 2015, tensions over the provision of L&D were at an all-time high, resulting in two diametrically opposed views dominating draft negotiating texts. The LDCs made political concessions by not including specific responsibilities of developed countries, but still requested compensation and liability. The developed

[79] Karen McNamara, 'Exploring Loss and Damage at the International Climate Change Talks' (2014) 5 *International Journal of Disaster Risk Science* 242; UNFCCC, *Report of the Conference of the Parties* (Footnote 73).

[80] Juan Hoffmaister et al., 'Warsaw International Mechanism for Loss and Damage: Moving from Polarizing Discussions Towards Addressing the Emerging Challenges Faced by Developing Countries' (Article, 6 January 2022) 2. https://www.twn.my/title2/climate/info.service/2014/cc140101/Opinion%20LD-060114.pdf.

[81] Liane Schalatek and Erin Roberts, 'Deferred Not Defeated: The Outcome on Loss and Damage Finance at COP26 and Next Steps', *Heinrich Boll Stiftung* (Web Page, 16 December 2021). https://us.boell.org/en/2021/12/16/deferred-not-defeated-outcome-loss-and-damage-finance-cop26-and-next-steps.

[82] Pekkarinen, Toussaint and van Asselt (Footnote 78) 33.

[83] Elisa Calliari, 'Loss and Damage: A Critical Discourse Analysis of Parties' Positions in Climate Change Negotiations' (2018) 21(6) *Journal of Risk* 725, 726.

[84] Hoffmaister et al. (Footnote 80) 2.

states, led again by the US, proposed text that deleted all references to L&D.[85]

During negotiations for the *Paris Agreement*, AOSIS proposed the creation either of a new international mechanism to address anthropogenic L&D, or a mandate to extend the existing functions of the *WIM* to give it more substantive heft.[86] Wealthy states spurned the offer as they were hoping to avoid discussion of the issue until 2016 when the *WIM* was due for a review. The US and its allies were fearful about adopting the proposal; they were concerned that too broad a mandate for the *WIM* would give rise to financial remedies through international litigation.[87]

The outcome of negotiations on the topic was the inclusion of Article 8 in the *Paris Agreement* requiring that, 'parties recognise the importance of averting, minimising and addressing L&D associated with climate change, including extreme weather and slow-onset events, and the role of sustainable development in addressing L&D'.[88] The article also extends the mandate of the *WIM* asking parties to enhance understanding, action and support, by utilising the *WIM* instrument.[89] However, problematically, no elaboration is provided to demonstrate how states should do that.[90] The deliberately vague language ensures that developed states are not bound to any commitments to mitigate L&D nor are they required to provide relief to vulnerable nations. Further, Article 8 expressly excludes any form of liability and compensation scheme being created.[91]

The addition of Article 8 to the *Paris Agreement* is considered by a number of scholars to have significantly advanced the momentum of L&D, giving the concept greater legitimacy within climate negotiations.[92]

[85] UNFCCC, *Work of the Contact Group on Item 3: Negotiating Text: Advance Unedited Version, 2nd session* (12 February 2015) 31.

[86] Linnéa Nordlander, Melanie Pill and Beatriz Martinez Romera, 'Insurance Schemes for Loss and Damage: Fools' Gold?' (2020) 20(6) *Climate Policy* 704, 705.

[87] Pekkarinen, Toussaint and van Asselt (Footnote 78) 33.

[88] *Paris Agreement Under the United Nations Framework Convention on Climate Change*, opened for signature 22 April 2016, TREATIES-XXVII.7.d (entered into force 4 November 2016) ('*Paris Agreement*'), art 8(1).

[89] Ibid., Article 8.

[90] Pekkarinen, Toussaint and van Asselt (Footnote 78) 34.

[91] *Paris Agreement* (Footnote 88) Article 8.

[92] Chad Boda et al., 'Framing Loss and Damage from Climate Change as the Failure of Sustainable Development' (2021) 13(8) *Climate and Development* 677.

Liao et al. argues that the inclusion of the provision puts a spotlight on the necessity to increase knowledge, action and funding for nations suffering from L&D.[93] However others, such as Wise, argue that any praise is exaggerated given the limitations of the vaguely written provision; the rejection of responsibility by developed states for climate losses and damages; and the exclusion of liability and compensation schemes in paragraph 51 of the agreement.[94]

2.3.3 Santiago Network

The Santiago Network on L&D was set up in Madrid, Spain (COP25) in 2019 to catalyse the technical assistance of organisations, bodies, networks and experts with a view to the implementation of approaches to L&D.[95] It has been rightly criticised as merely consisting of precatory language exhorting states to 'catalyse the technical assistance ... for the implementation of relevant approaches ... in developing countries that are particularly vulnerable to the adverse effects of climate change'.[96] States could not agree on the ambit of the network, ultimately merely providing finances for its operation, with states such as Germany pledging €10 million.[97]

[93] Liao et al. (Footnote 65).

[94] Jacob Wise, 'Climate Change Loss and Damage Litigation: Infeasible or a Useful Shadow?' (2021) 38(1) *Wisconsin International Law Journal* 687, 696; Bethany Tietjen and Tarun Gopalakrishnan, 'Loss and Damage Funding in the UN Climate Negotiations: From Dialogue to Reality' (2023) 65(3) *Environment: Science and Policy for Sustainable Development* 18, 23; UNFCCC, 'The Paris Outcome on Loss and Damage (Article 8 of the Paris Agreement and Decision 1/CP.21 Paragraphs 48-52 (FCCC/CP/2015/L.9/Rev.1.)' (2015). https://unfccc.int/files/adaptation/groups_committees/loss_and_damage_executive_committee/application/pdf/ref_8_decision_xcp.21.pdf.

[95] UNFCCC, 'About the Santiago Network', *United Nations Climate Change* (2023) (Web Page). https://unfccc.int/santiago-network/about.

[96] Jacob Katz Cogan, 'Contemporary Practice of the United States Relating to International Law' (2023) 117 (2) *The American Journal of International Law* 330, 331 332 citing *Warsaw International Mechanism for Loss and Damage Associated with Climate Change Impacts and its 2019 Review*, para. 43, Decision 2/CMA.2 in *Report of the Conference of the Parties Serving as the Meeting of the Parties to the Paris Agreement on Its Second Session*, Held in Madrid from 2 to 15 December 2019, UN Doc. FCCC/PA/CMA/2019/6/Add.1. https://unfccc.int/sites/default/files/resource/cma2019_06_add.01_.pdf.

[97] Schalatek and Roberts (Footnote 81).

2.3.4 *Glasgow Dialogue on Finance for Loss and Damage and Glasgow Climate Pact*

One of the first significant signs of tangible progress on the L&D issue occurred in 2021 at COP26 in Glasgow, when host Scotland became the first developed country to recognise L&D, pledging US$2.3 million towards the funding.[98] However, US Special Presidential Envoy for Climate John Kerry made it clear that the US would never establish any legal structure that was tied to compensation or liability by it and other developed states.[99]

Despite such rhetoric, the Glasgow Facility on Loss and Damage Finance was proposed at the meeting to serve as a process to establish a facility for financing; it was backed by 134 countries of the G77 and the China coalition. After torrid negotiations, the facility was not accepted but a compromise Glasgow Dialogue on Finance for Loss and Damage was created.[100] The 'Glasgow Dialogue' is to run until June 2024 with the aim of creating a conversation to 'discuss the arrangement for the funding of activities to avert, minimise and address loss and damage associated with the adverse impacts of climate change'.[101] At the same meeting, the Glasgow Climate Pact was agreed to specifically mention L&D; it presses developed countries and applicable organisations to render 'enhanced and additional support' for L&D-linked actions.[102]

[98] David South, 'COP27 Loss and Damage Fund—Financial Band Aid or Effective Mechanism?' (2023) 39(11) *Climate and Energy* 29, 30.

[99] Saeed Shah and Matthew Dalton, 'COP27 Talks Weigh Who Should Pay for Climate Damage to Poor Countries', *The Wall Street Journal Eastern Edition* (Online, 16 November 2022). https://www.wsj.com/articles/cop27-talks-weigh-who-should-pay-for-climate-damage-to-poor-countries-11668598218.

[100] Liane Schalatek, '"Glass Less Than Half Full"—Glasgow Climate Finance Outcomes Leave Much Room for Improvements Despite Some Wins', *Heinrich Boll Stiftung* (Web Page, 15 December 2021). https://us.boell.org/en/2021/12/15/glass-less-half-full-glasgow-climate-finance-outcomes-leave-much-room-improvements#5B ('Glass Less than Half Full').

[101] Liao et al. (Footnote 65).

[102] Ibid.

2.4 COP27 and a New Framework

Ahead of the COP27 meeting, COP President Sameh Shoukry assured developed states that talks of any L&D fund would not involve liability—in order to gain their participation in the upcoming talks on the issue.[103] COP27 was held in November 2022 to consolidate intergovernmental efforts to combat climate change.[104] At the meeting, Egypt's role as host and chair facilitated a stronger focus on the needs of LDCs. The Egyptian leadership chose to highlight L&D as a key focus of the summit with strong emphasis on natural disasters occurring in the Middle East region, positioning L&D negotiations at the forefront of discussions.[105]

Developing practical strategies for implementing an L&D mechanism in response to the climate crisis was a key focus of opening statements made at COP27.[106] Australia, representing the Umbrella Group,[107] emphasised the goals of 'operationalizing the Santiago Network on loss and damage' to support such an L&D apparatus.[108] Antigua and Barbuda, speaking on behalf of AOSIS, stated that while progressing the L&D agenda item into a tangible project would require intensive 'collaborative action', an L&D mechanism could operate to preserve the 1.5 °C goal of the *Paris Agreement*.[109] Senegal, on behalf of the LDCs, acknowledged the importance of concentrating on L&D at COP27, noting the

[103] Shah and Dalton (Footnote 99).

[104] World Wildlife Fund, 'It's Time to Shift from Promises to Action: The UN Climate Change Conference, COP27, Was Held in Sharm El-Sheikh, Egypt from 6 to 18 November 2022'. https://wwf.panda.org/discover/our_focus/climate_and_energy_practice/cop27/.

[105] Tietjen and Gopalakrishnan (Footnote 94) 19.

[106] 'COP27 Highlights: Sunday, 6 November 2022', *Earth Negotiations Bulletin* 1. https://enb.iisd.org/sites/default/files/2022-11/enb12807e.pdf.

[107] The Umbrella Group is a coalition of COP member-states that formed following the adoption of the *Kyoto Protocol*. The group currently consists of Australia, Canada, Iceland, Israel, Japan, New Zealand, Kazakhstan, Norway, Ukraine and the US. https://unfccc.int/process-and-meetings/parties-non-party-stakeholders/parties/party-groupings#:~:text=They%20were%20the%20first%20to%20propose%20a%20draft,formed%20following%20the%20adoption%20of%20the%20Kyoto%20Protocol.

[108] 'COP27 Highlights: Sunday, 6 November 2022' (Footnote 106) 1.

[109] Ibid.

requirement to focus not only on adaptation, but on L&D, an urgent matter for those 'suffering from the climate crisis'.[110]

Several non-governmental organisations (NGOs) in attendance also argued for an L&D-oriented climate change solution. While the Climate Action Network generally acknowledged, 'the global climate crisis and the loss and damage it causes', Demand Climate Justice (approaching the matter more specifically) called for 'a loss and damage funding facility' based on 'public funds, just transitions, and a conflict-of-interest policy for the *UNFCCC*'s process'.[111]

Seeking to resolve the L&D issues, the COP agreed that informal consultations would convene on agenda item 8f, namely: 'Matters relating to funding arrangements responding to loss and damage associated with the adverse effects of climate change, including a focus on addressing loss and damage'.[112] A joint contact group was formed on 6 November 2022 to deal with this specific point of negotiation.[113]

From the outset, the main point of contention for the joint contact group was 'a call for a new funding facility with dedicated resources for loss and damage'.[114] While parties' support for a new dedicated L&D fund was widespread, developing countries—especially SIDS and LDCs—pushed for a firm commitment at COP27 to establish and operationalise an L&D fund as soon as practicable, calling for significant new and additional resources to be applied to the fund.[115] An unnamed head of government also called for a 'fit for purpose' multilateral L&D fund that would be operational by 2024.[116]

Parties' support also extended to proposing how the L&D fund could be financially sourced. At the high-level segment on 8 November 2022, the prime minister of Antigua and Barbuda emphasised that fossil fuel

[110] Ibid.

[111] Ibid. 2.

[112] 'COP27', *United Nations Climate Change* (Web Page). https://unfccc.int/event/cop-27?item=8. The items under agenda item 8 generally dealt with the topic of long-term finance, although this summary focusses on 8f as the sub-item that dealt with loss and damage specifically.

[113] 'COP27 Highlights: Sunday, 6 November 2022' (Footnote 106) 3.

[114] IISD, *Earth Negotiations Bulletin*, Vol 12, No. 818, 23 November 2022. https://enb.iisd.org/sites/default/files/2022-12/enb12818e_0.pdf.

[115] Ibid. 'Summary Report', 9.

[116] 'COP27 Highlights: Sunday, 6 November 2022' (Footnote 106) 2.

companies must pay a global carbon tax for L&D, in addition to calling unequivocally for the fund to be operational by 2024.[117] Switzerland, on behalf of the Environmental Integrity Group, also proposed a broad source of funding for an L&D fund, arguing that 'flexibility' on L&D finance is contingent on a 'strong outcome on mitigation and that contributions to a fund would come from a broadened donor base of public and private resources and all countries with a capacity to provide support'.[118] Several developing countries at COP27 advocated establishing this funding facility and support was also received from developed nations for any L&D fund to be operationalised by 2024.[119]

Despite this popular support for a new fund, there were various forms of opposition, typically in the form of how to address the technical complexities of such an entity. Notwithstanding developed nations' support for establishing a dedicated fund, most adopted the alternative view that existing financial institutions were a better vehicle to distribute funds. For example, in lieu of an L&D fund, it was suggested that a funding 'window' might be created within the *UNFCCC* financial mechanism.[120] However, in a response to the *Report of the Green Climate Fund*, developing states noted (in the context of its utility) the need for 'more substantive and prescriptive' access to funds, which '[covered] a broader range of activities' such as inter alia L&D, suggesting technical issues with locating such a *UNFCCC* funding 'window'.[121]

Further, raising a technical constraint but not an objection to an L&D fund per se, it was noted that any funding solution to the climate crisis must respect the principles of the *UNFCCC*. During the stocktaking plenary, Bolivia, Saudi Arabia (on behalf of the Arab Group) and China directly 'underscored the need to respect the principles of the Paris Agreement' and the *UNFCCC*, regarding any decisions related to deriving

[117] Ibid.

[118] Ibid.

[119] On 12 November, several developing country groups reiterated their call to establish a finance facility for L&D at COP27/CMA4 and set out a clear roadmap to ensure its full operationalisation by 2024. 'COP27 Highlights: Saturday, 12 November 2022', *Earth Negotiations Bulletin* 2. https://enb.iisd.org/sites/default/files/2022-12/enb12818e_0.pdf.

[120] UNFCC (Footnote 33), art 11.

[121] 'Summary Report' (Footnote 115) 9.

funding for L&D finance projects.[122] Zambia, on behalf of the African Group, also asserted that, 'any [L&D funding] package must include the [Global Goal on Adaptation]' under the *Paris Agreement* as the counterpart to its 1.5 °C goal.[123]

Parties' views also diverged on the preferred type of funding arrangement. Several developing country groups advocated a standalone funding facility.[124] Other developed nations suggested alternatives such as a dedicated L&D window under the Green Climate Fund, Global Environment Facility (GEF) or Adaptation Fund, consistent with a preference to utilise pre-existing *United Nations Environment Programme* finance mechanisms.[125] Other funding mechanisms (risk insurance facilities and bilateral support) were also proposed by developed nations to shape the L&D fund structure.[126]

The potential role of pre-existing mechanisms in facilitating the operation of this L&D fund was also discussed. For example, a multifaceted macro-organisational role for the Glasgow Dialogue was proposed by various developed countries including assisting with the current global L&D situation; identifying how appropriate institutions can provide support; coordinating global actors; and pointing to gaps in (and mapping potential sources of) funding.[127] Micro strategies to assist with the fund's operation through workshops, technical papers, regional meetings, receipt of submissions and production of annual reports were also proposed.[128] In particular, and consistent with these functions, the Glasgow Dialogue was envisioned to 'provide the space to discuss specific issues, including

[122] 'COP27 Highlights: Thursday, 17 November 2022', *Earth Negotiations Bulletin* 2. https://enb.iisd.org/sites/default/files/2022-12/enb12818e_0.pdf.

[123] Ibid.

[124] 'COP27 Highlights: Thursday, 10 November 2022', *Earth Negotiations Bulletin* 1. https://enb.iisd.org/sites/default/files/2022-12/enb12818e_0.pdf.

[125] Ibid.

[126] Ibid.

[127] 'COP27 Highlights: Friday, 11 November 2022', *Earth Negotiations Bulletin* 1. Several of these functions were subsumed into the role of the transitional committee (see section). https://enb.iisd.org/sites/default/files/2022-12/enb12818e_0.pdf. See ENB, 'Summary Report' (Footnote 115) 9.

[128] 'COP27 Highlights: Friday, 11 November 2022' (Footnote 127) 1.

slow-onset events, rapid response, the role of MDBs [multilateral development banks] and debt relief'.[129] However, it was noted that this portfolio was at odds with the highly ambitious target of an operational L&D fund by 2024.[130]

Additionally, despite operational aspects of an L&D fund not being settled at COP27, FCCC/CP/2022/L.14 states that under the *WIM*, states should 'continue exploring further opportunities and modalities for engaging national stakeholders, including loss and damage contact points and national focal points and collaborating ... under and outside the [*UNFCCC*] and Paris Agreement'.[131]

Negotiations also dealt with whether the Santiago Network should coordinate national L&D contact points in support of an L&D fund operation. It was determined that the Santiago Network would be restructured 'to provide guidance and oversight ... on the effective implementation of the functions of the network', covering 'topics relevant to averting, minimizing and addressing [L&D]'.[132]

The focus on reaching tangible decisions regarding the structure and composition of an L&D fund was prevalent in the final stages of COP27 negotiations, consistent with the Maldives defining success at COP27 as a cover decision that, inter alia, established an L&D funding system.[133] At the stocktaking plenary convened on 17 November 2022, dealing with a draft text, 'based on multiple bilateral consultations', the *UNFCCC* executive secretary advocated such 'tangible outcomes' with 'clear timelines and milestones'.[134]

The European Union (EU) and other developed states proposed the establishment of an L&D fund under the Meeting of the Parties to the Paris Agreement (CMA), 'for [those] most vulnerable countries', and that a 'workplan [would] be agreed upon by the CMA'.[135] Similarly, Colombia, on behalf of the Independent Alliance of Latin America and the Caribbean called for the EU's proposal on debt relief that it viewed

[129] Ibid.

[130] Ibid.

[131] 'Summary Report' (Footnote 115) 9.

[132] Ibid.

[133] 'COP27 Highlights: Saturday, 12 November 2022' (Footnote 119).

[134] 'COP27 Highlights: Thursday, 17 November 2022' (Footnote 122).

[135] Ibid.

as crucial to helping create fiscal space for countries to undertake climate action.[136]

At an informal evening plenary, many LDCs agreed with the EU proposal to establish an L&D facility for the most vulnerable nations, with others outlining the need for a 'broad donor base' to support such a facility.[137] This is also consistent with the Zambian proposal, on behalf of the African Group, that COP27 supports a decision establishing a new operating entity of the *UNFCCC* financial mechanism. Ghana, on behalf of the Climate Vulnerable Forum, was similarly proactive in '[urging]' important developments; 'the creation of a loss and damage finance facility under the COP and CMA; clear commitments for L&D funding; and resources from developed countries and others in a position to provide support'.[138]

However, Pakistan, on behalf of the G77/China acknowledged that, 'there [were] still many divergent views and unanswered questions on loss and damage finance', with Antigua and Barbuda for AOSIS, similarly, '[expressing] concern over how the [L&D] financing [arrangements]' were to be managed.[139] On the precipice of L&D negotiations again failing, negotiations were resolved on 19 November 2022 by agreement from the EU and other developed nations to create an L&D fund structure, broadly based on climate mitigation and with the goal of it helping to facilitate the goal of limiting global temperature increases to 1.5 °C. Ultimately, it was unrelenting pressure by LDCs to create the fund that caused the EU to agree to its establishment with conditions, followed by a last-minute agreement by the US negotiators.[140]

The main operative decisions from these negotiations are summarised below:

- To establish new funding arrangements for assisting developing countries that are particularly vulnerable to the adverse effects of

[136] Ibid.

[137] Ibid.

[138] 'COP27 Highlights: Thursday, 17 November 2022' (Footnote 122).

[139] Ibid.

[140] Shah and Dalton (Footnote 99).

climate change in response to L&D. And that these new arrangements complement and include sources, funds, processes and initiatives under and outside the Convention and Paris Agreement;

- To establish a fund for responding to L&D whose mandate includes a focus on addressing L&D.[141]

In announcing the decision, COP President Sameh Shoukry noted that the L&D fund raised several issues of liability and compensation for developed countries with respect to historic emissions.[142] This caveat mirrors the language of L&D in both Article 8 and the cover decision of the *Paris Agreement*. This decision was arguably a compromise to ensure that developed countries would approve the LDF. Thus, the LDF serves as a tool for risk management and restorative justice rather than a tool for punishment.[143]

2.5 The Transitional Committee

A Transitional Committee was tasked with making recommendations on the structure and operation of an L&D fund at COP28, including who will receive funding, sources of funding and how the fund will work in relation to existing climate funds.[144] The committee included 14 members from developing countries and 10 members from developed countries.[145] Notably, Pacific SIDS did not have representation on the committee despite bearing the most acute effects of climate change.

The committee held its second meeting in Bonn on 25–27 May 2023. Members discussed various aspects of an L&D fund, such as institutions

[141] Ibid.

[142] Jacob Cogan, 'The United States Agrees to Loss and Damage Fund at COP27' (2023) 117(2) *American Journal of International Law* 330, 334–335.

[143] Mahase (Footnote 19) 1.

[144] 'Transitional Committee', *United Nations Climate Change* (Web Page, July 19, 2023). https://unfccc.int/topics/adaptation-and-resilience/groups-committees/transitional-committee; 'COP27 Reaches Breakthrough Agreement on New "Loss and Damage" Fund for Vulnerable Countries', *United Nations Climate Change* (Web Page, 20 November 2022). https://unfccc.int/news/cop27-reaches-breakthrough-agreement-on-new-loss-and-damage-fund-for-vulnerable-countries.

[145] 'Membership—Transitional Committee', *United Nations Climate Change* (Web Page, 26 May 2023). https://unfccc.int/tc-membership.

that might facilitate new funding arrangements, and potential sources of necessary finance.[146] The Santiago Network was seen by delegates as an important part of the funding arrangements for an L&D fund to ensure technical assistance and capacity building, similar to the readiness programme of the Green Climate Fund.[147]

A second workshop took place in a hybrid format on 15–16 July 2023, with an in-person component taking place in Bangkok, Thailand, with over 250 participants attending. Attendees discussed among other matters potential advantages and disadvantages regarding the placement of the fund: whether it should be a standalone fund; a new fund housed in an existing institution under the *UNFCCC*; or a new fund housed in an existing institution outside the *UNFCCC*. Potential options included a new Adaptation Fund; the Global Climate Fund, the GEF and the World Bank.[148]

Delegates also proffered a range of suggestions on potential thematic priorities for an L&D fund. For example, should it:

> support actions to address slow-onset events, and take an additive approach to support to complement or supplement existing support available for extreme weather events … [or focus] on extreme weather events [that] can target particular phases of the post-event response (e.g. one week post-immediate response/humanitarian interventions and longer-term rehabilitation and reconstruction)'.[149]

However, no decisions were made at the meeting, with attendees instead choosing to defer this to COP28.

[146] 'Transitional Committee on Loss and Damage Holds Second Meeting in Bonn', *United Nations Climate Change* (Web Page, 1 June 2023). https://unfccc.int/news/transitional-committee-on-loss-and-damage-holds-second-meeting-in-bonn.

[147] Ibid.

[148] Third meeting of the Transitional Committee, Report of the second workshop on addressing loss and damage in the context of decisions 2/CP.27 and 2/CMA.4, TC3/2023/4, 17 August 2023, 2–3, 12–14. https://unfccc.int/sites/default/files/resource/TC3_WS2_SummaryReport.pdf.

[149] Ibid. 9.

2.6 Outstanding Issues to be Resolved Going Forward

The mere fact that a dedicated L&D fund, 30 years in the making, was created at COP28 is a historic step towards achieving climate justice.[150] As Pakistan's former Foreign Minister Bilawal Bhutto Zardari wrote: 'while the agreement does not establish the legal responsibility of those who have contributed the most to climate change and global warming, it does confirm the central principle of climate justice'.[151] The chair of AOSIS states its creation means that, 'the international community has restored global faith in this critical process that is dedicated to ensuring no one is left behind'.[152] The creation of a fund also refutes the idea that arguments for action based purely on climate justice and 'moral persuasion' are insufficient to progress the issue.[153] However, a number of key issues remain to be resolved if an L&D fund is to become both effective and a vehicle to promote climate justice. If the history of these negotiations is any guide, these issues will not be easy to resolve.[154]

Mahase points to a major criticism from the climate science community that L&D payouts 'could take decades' because of arguments over 'contested science' that may derail the operation of the fund.[155] This is further rendered problematic by the controversial definition of 'climate change' adopted under the *UNFCCC*, which requires that proof of L&D requires '[attribution] (of) whatever portion of losses have occurred to the emission of greenhouse gases'.[156] In practice, because of the inherent scientific

[150] Raymond Clemencon, '30 Years of International Climate Negotiations: Are They Still Our Best Hope?' (2023) 32(2) *The Journal of Environment & Development* 114, 116.

[151] Zardari (Footnote 28).

[152] Wyns (Footnote 20) 1.

[153] Tikhomirova (Footnote 25) 38.

[154] Michael Franczak, 'Options for a Loss and Damage Financial Mechanism', *International Peace Institute* (Web Page, 14 October 2022). https://www.ipinst.org/2022/10/options-for-a-loss-and-damage-financial-mechanism.

[155] Elisabeth Mahase, 'Climate Change: "Loss and Damage" Fund Payouts Could Take Decades, Scientists Warn' (2022) 379 *BMJ*. https://www.bmj.com/content/379/bmj.o3050.

[156] Ibid.

uncertainty of this threshold, Roger Piekle Jr. argues that this requirement potentially renders L&D compensation virtually infeasible.[157]

Masood argues trenchantly that while agreement to an L&D mechanism is a positive step for global climate change policy, a 'failure to rein in fossil-fuel interests' not only facilitates their continued production, but could also nullify the success of an L&D fund in actually addressing climate-related harm.[158] The executive vice president of the EU further noted that COP27's L&D agreement was 'not enough of a step forward for people and planet', in that, '[i]t does not bring enough added efforts from major emitters to increase and accelerate their emissions cuts'.[159]

Conceptually, climate compensation mechanisms are technically problematic, particularly—according to Pill—in an L&D context.[160] L&D from climate change is particularly 'challenging to measure', partly because of 'specificity and temporal and spatial scales', in the sense that ecological harm tends to transpire over several decades or is ongoing years after a climate-adverse event.[161] It is therefore clear that there are immediate design and implementation matters to be resolved for an L&D fund to be operational. Wyns notes several of these questions: 'How will the fund be structured? Who will pay into it? How much will be paid? Who will manage it?'[162] Added to this are questions such as whether developed states are under an *obligation* to contribute to the fund because of their historic responsibility of emissions, or whether any contribution is *voluntary*.[163] Also, will the fund apply to all LDCs who apply or merely those

[157] Ibid.

[158] Ehsan Masood et al., 'COP27 Climate Talks: What Succeeded, What Failed and What's Next', *Nature* (21 November 2022). https://www.nature.com/articles/d41586-022-03807-0.

[159] Aljazeera, 'Not Enough for People and Planet: Critics Denounce COP27 Deal', 20 November 2022. https://www.aljazeera.com/news/2022/11/20/not-enough-for-people-and-planet-critics-denounce-cop27-deal.

[160] Melanie Pill, 'Towards a Funding Mechanism for Loss and Damage from Climate Change Impacts' (2022) 35 *Climate Risk Management*. https://www.sciencedirect.com/science/article/pii/S2212096321001224.

[161] Emily Boyd et al., 'Loss and Damage' (2021) 4(10) *OneEarth*. https://www.sciencedirect.com/science/article/pii/S2590332221005376.

[162] Wyns (Footnote 20).

[163] Cogan (Footnote 96) 334–335.

that are *particularly vulnerable*?[164] How will the amounts to be allocated to states be calculated?

The COP27 text outlined that an L&D fund should assist developing countries that are 'particularly vulnerable' to the adverse effects of climate change. However, beyond this 'broad eligibility criterion', which countries qualify for compensation remains ambiguous.[165] How will L&D be accurately assessed and by what methodologies?[166] Will payments be monitored and subject to conditions, or will states have complete autonomy as to the use of funds? Will funds be dispersed as block grants to states or will they be project specific? The final text of COP27 neglects to answer these questions but does state that any funds should be new.[167] Therefore, it will be a challenge to meet the funding targets set.

One key justice issue that will need to be resolved is that of attribution of responsibility, which is difficult to determine where multiple actors across several generations have generated GHG emissions. Though developed states have benefitted from the industrialisation process on which their economies were built, the idea of holding current and/or future generations responsible for the actions of previous generations, in which they had no say, is controversial.[168]

Further, it could be argued that even previous generations bear no culpability given that there was no scientific understanding or consensus on the harmfulness of emissions at the time of their production.[169] Also, as GHGs can be generated by both natural and human processes, the causal pathway typically required by legal approaches such as negligence when attributing responsibility is unclear.[170] Last, attributional science is a developing field that has not yet progressed to the point where it

[164] Ibid.

[165] Olivia Sendecznt, Manjeet Dhakal and Sneha Pandey, '2023 Will Shape the Loss and Damage Fund for Years to Come—Have Your Say Now', *Climate Analytics* (Blog Post, 9 January 2023). https://climateanalytics.org/blog/2023/2023-will-shape-the-loss-and-damage-fund-for-years-to-come-have-your-say-now/.

[166] Cogan (Footnote 96) 334–335.

[167] Clemencon (Footnote 150) 124.

[168] Craig A Johnson, 'Holding Polluting Countries to Account for Climate Change: Is Loss and Damage up to the Task?' (2017) 34(1) *The Review of Policy Research* 54.

[169] Mechler et al. (Footnote 58) 45.

[170] Garcia (Footnote 11) 192–193.

is possible to definitively say whether a certain GHG emitter bears sole responsibility for causing climate change.[171]

Inherently problematic is the difficulty in distinguishing the impacts of anthropogenic climate change from other causes of harm in the context of both slow-onset climatic events and rapid-onset weather events. However, this legal difficulty does not mean that developed states cannot agree to be 'liable' for their historical GHG emissions. The problem will always be the amount, time period and apportioning of responsibility to different developed states.

A key obstacle to be overcome is the quantum required for 'total adaptation', estimated at US$2.5 trillion.[172] Wyns emphasises not only the requisite scale of finance required, but the 'speed with which it can be delivered', as key factors to consider for any fund design.[173] Further historically, climate finance institutions struggle to effectively distribute funds to those most vulnerable.[174] Wyns notes that not only do existing climate finance entities such as the Green Climate Fund 'take years to distribute funds', but that when funds are distributed, they 'often struggle to reach local communities and the most vulnerable'.[175]

On this basis, Wyns emphasises that the COP27 L&D fund 'should have a greater emphasis on last mile delivery of support to communities in need'.[176] Both of these practical concerns echo comments of Jean Su of the Centre for Biological Diversity, that there remain 'big questions' regarding the function of the fund, including 'whether finance will actually get delivered' to vulnerable states.[177]

171 Mechler (Footnote 58) 115.

172 Robin McKie, 'Five Crucial Issues in Fight to Save Planet—And What COP27 Did About Them', *The Guardian* (20 November 2022). https://www.theguardian.com/environment/2022/nov/19/five-crucial-issues-in-fight-to-save-planet-and-what-cop27-did-about-them#:~:text=Loss%20and%20damage,suffering%20most%20from%20climate%20change; Wyns (Footnote 20).

173 Wyns (Footnote 20).

174 See generally Wyns (Footnote 20); Pill (Footnote 160).

175 Wyns (Footnote 20).

176 Ibid.

177 Rachel Frazin and Sharon Udasin, 'Uncertainties About Climate Compensation Fund Trigger Scepticism', *The Hill* (22 November 2022). https://thehill.com/policy/energy-environment/3745628-uncertainties-about-climate-compensation-fund-trigger-scepticism/.

Regarding the different forms of justice, an L&D fund would promote procedural justice because it provides for 'meaningful inclusion and participation of those most affected by climate change at all levels of decision-making'.[178] Previously, the issue had been caught up in what Roberts and Parks argue was a 'North–South stalemate on climate policy' that had developed because of 'rich donor countries ... persuad[ing] recipient countries ... to take the environmental actions of [lowest] priority to them'.[179] The decision to provide compensation has the potential to break the deadlock if implemented effectively and equitably, thus improving good will between the two factions, which might lead to breakthroughs in regard to other climate-related issues.

The fund also potentially facilitates distributive justice because its raison d'etre is to provide finance to those peoples most vulnerable to impacts from climate change. However, given that the details of the fund have not been clarified, any positive impacts regarding its equitable distributive nature are at best speculative. The Stockholm Environment Institute states it allows LDCs to seek restorative justice,[180] and without this, 'we cannot move forward on an equal, level playing field'.[181] The Stockholm Environment Institute further argues that restorative justice will only truly be achieved if the recipients of finance are meaningfully included in the decision-making process. However, given the fund's nascent stage, we cannot state to what extent LDCs will be part of the decision-making process.

The wariness of developed countries in broaching any topic related to compensation or liability continues to present the greatest barrier to further progress on L&D action. Reluctance from developed states to include liability language in climate texts stems from the fear of exposure to unlimited financial liability.[182] Developed states have also been reluctant to assent to a fund for over 30 years, being concerned about

[178] Stockholm Environment Institute, *Operationalizing Finance* (Footnote 23) 11.

[179] J Timmons Roberts and Bradley C Parks, *A Climate of Injustice: Global Inequality, North–South Politics, and Climate Policy* (MIT Press, 2007) 6.

[180] Restorative justice is defined as, 'the restoration of dignity, agency and capabilities to those who have lost it'; Stockholm Environment Institute, *Operationalizing Finance* (Footnote 23).

[181] Elizabeth Mahase, 'Climate Change: Loss and Damage Fund Payouts Could Take Decades, Scientists Warn' (2022) 379 *BMJ (Online)*.

[182] Damian Carrington, 'What Is "Loss and Damage" and Why Is It Critical for Success at COP26?', *The Guardian* (Online, 14 November 2021). https://www.theguardian.com/environment/2021/nov/13/what-is-loss-and-damage-and-why-is-it-critical-for-success-at-cop26.

not knowing the extent of liability to which they are exposed—at law described as opening the 'floodgates' to a plethora of litigation.[183]

However, as noted by Mrinalini Shinde, a former legal advisor to the *UNFCCC*, 'For the fund to be truly transformative, it is important that it is driven by the motivation to set right historical wrongs'.[184] As Perry argues, it is essential that developed states do not view their contributions to the fund as acts of charity or good will, but rather just compensation for societies that were not only marginalised and colonised by developed nations but also have benefitted less from carbon usage.[185]

To better embed and potentially operationalise the issue of climate justice, states can further develop the framework of 'common but differentiated responsibility' ('CBDR'), which has underpinned decision-making in international climate negotiations. CBDR was first formulated as Principle 7 of the *Rio Declaration on Environment and Development* in 1992.[186] It represents an international consensus that the ongoing responsibility to protect our planet from anthropogenic climate change ought to be shared among all nation states, but not necessarily *equally* shared. It accepts that developed states bear a greater responsibility to address climate change because of their greater contribution to the earth's CO_2 emissions, and their greater financial and technological ability to act. However, the precise interpretation of this principle remains widely contested, including how the responsibilities of addressing climate change ought to be distributed in a manner that remains fair to all parties involved.[187]

The principle of CBDR is particularly relevant with regard to L&D. Developing states have contributed the least to ongoing climate change and have the fewest resources, yet are disproportionately affected by

[183] 'See You in Court: How Climate Lawsuits Could Sharpen COP27 Loss and Damage Talks', *Climate Home News* (Web Page, 22 November 2022). https://www.climatechangenews.com/2022/11/02/see-you-in-court-how-climate-lawsuits-could-sharpen-cop27-loss-and-damage-talks/.

[184] Wyns (Footnote 20).

[185] Keston K Perry, *Realising Climate Reparations: Towards a Global Climate Stabilization Fund and Resilience Fund Programme for Loss and Damage in Marginalised and Former Colonised Societies* (Report, 1 March 2020) 20. https://ssrn.com/abstract=3561121.

[186] Jonathan Boston, Andrew Bradstock, and David Eng, *Public Policy: Why Ethics Matters* (The Australian National University E Press, 2010) 141.

[187] Ibid. 141–142.

climate change and the resulting L&D.[188] Meanwhile, developed countries have the highest global GHG emissions, the most advanced economies and greater capabilities to address climate change; thus, it is argued that they bear more responsibility to do so.[189]

Climate justice therefore recognises that developed nations that have benefitted from carbon-intensive development that has led to global warming and climate change should bear the costs of L&D experienced by developing countries that have not contributed to GHG emissions to anywhere near the same extent.[190] The *UNFCCC* has positioned 'equity' as a pillar of its processes, and the concept of 'climate justice' is referred to explicitly in the *Paris Agreement* Preamble. Problematically, the Preamble does not explicitly detail how to achieve climate justice, leaving the burden to academics and NGOs to further define the concept and how to effectively operationalise it.[191]

By setting up an L&D fund, developed states are conceding that while all states have shared responsibility to address environmental destruction, developed states that have been greater contributors to the cause of the climate crisis need to help bear the costs that vulnerable countries will experience. The fund also appears to be acknowledging the international environmental law principle of 'polluter pays'. This principle requires that the costs of pollution are borne by those responsible for causing the pollution.[192] Various 'polluter pays oriented options have been proposed

[188] IPCC, *Climate Change 2022: Impacts, Adaptation and Vulnerability*. Contribution of Working Group II to the Sixth Assessment Report of the Intergovernmental Panel on Climate Change (Report, 2022) [1.4.4.2].

[189] United Nations Environment Programme, *Bridging the Gap—Enhancing Mitigation Ambition and Action at G20 Level and Globally* (Report, 2019) 8. 21 September 2018. https://www.unep.org/resources/bridging-gap-enhancing-mitigation-ambition-and-action-g20-level-and-globally.

[190] Sam Adelman, 'Climate Justice, Loss and Damage and Compensation for Small Island Developing States' (2016) 7(1) *Journal of Human Rights and the Environment* 32, 36.

[191] Richard Kinley et al., 'Beyond Good Intentions, to Urgent Action: Former UNFCCC Leaders Take Stock of Thirty Years of International Climate Change Negotiations' (2021) 21(5) *Climate Policy* 593, 597–598; Stockholm Environment Institute, *Operationalizing Finance* (Footnote 23) 11.

[192] Jose Felix Pinto-Bazurco, *How to Enforce the Polluter-Pays Principle* (Policy Brief 31, International Institute for Sustainable Development, February 2022). https://www.iisd.org/system/files/2022-02/still-one-earth-polluter-pays-principle.pdf.

to source funding including taxing fossil fuel and other carbon intensive industries'.[193]

2.7 Conclusion

The decision to establish a dedicated L&D fund marks a historic breakthrough in climate change history as it follows widespread pressure from developing nations to compensate them for the harm caused by wealthy nations.[194] Historically, developed states, particularly the US, EU, Russia, China, Saudi Arabia and other Organization of the Petroleum Exporting Countries have resisted compulsory or defined commitments and instead sought voluntary and minimal contributions towards climate L&D that minimises potential future liabilities.[195] However, at COP27, the EU and US showed a willingness for a climate fund to be established—which, as demonstrated in the previous section—can be to the benefit of all parties moving forward.

The COP27 meeting has been called a 'step towards justice'; however, it is a very small step and it remains unclear at this juncture how an L&D fund will be effectively implemented and administered, particularly in the absence of an agreed-to definition.[196] Currently, all of the legal instruments created that empower L&D action exclude questions of liability and compensation.[197] Creation of an L&D fund with the purpose of providing support and assistance for vulnerable countries demonstrates a first move by the global community towards recognising equity and climate justice.[198] SIDs, LDCs and climate activists have argued that

[193] Ibid.

[194] 'COP27 Reaches Breakthrough Agreement' (Footnote 144).

[195] Kate Abnett and Dominic Evans, 'Explainer: Who Will Pay for Climate "Loss And Damage"?', *Reuters* (Online, 22 November 2022). https://www.reuters.com/business/cop/who-will-pay-climate-loss-damage-2022-11-20/.

[196] 'COP Closes with Deal on Loss and Damage: "A Step Towards Justice" Says UN Chief', *UN News: Global Perspective Human Stories* (Web Page, 22 November 2022). https://news.un.org/en/story/2022/11/1130832.

[197] 'See You in Court' (Footnote 183).

[198] Kinley et al. (Footnote 191) 598.

effective climate justice will not be achieved until those responsible for climate change are held morally and legally accountable.[199]

However, at COP27 the concerns of developed states on these points were alleviated, with a pre-agreement that outcomes would 'not involve liability or compensation', allowing them to engage in negotiations without concern for their legal liability.[200] Thus, no real shift has occurred in terms of developed states taking legal responsibility for their past behaviour, demonstrating that there is still work to be done towards fully integrating an effective climate justice template.[201]

Developing states maintain that the concept of L&D is linked to climate justice, whereas developed nations have relegated it to an issue of mere risk management.[202] Without any explicit reference to liability in the legal framework, it is unclear how developed states are going to be held accountable, rendering the impact of an L&D fund's creation as potentially tokenistic in nature.[203] An L&D provision will not necessarily change the trajectory of the climate crisis. However, its successful implementation would be life-changing for the population of vulnerable communities impacted both presently and, in the future, and that is a goal worth striving for.

[199] Zardari (Footnote 28) 1; Angus Naylor and James Ford, 'Vulnerability and Loss and Damage Following the COP27 of the UN Framework Convention on Climate Change' (2023) 23(38) *Regional Environmental Change* 37, 38.

[200] Kate Raffety et al., *What Happened at COP 27 on Loss and Damage and What Comes Next?* (Report, Loss and Damage Collaboration, 20 November 2022) 5; South (Footnote 98) 31; Cogan (Footnote 142) 333.

[201] Zardari (Footnote 28) 1; Naylor and Ford (Footnote 199) 38.

[202] Tietjen and Gopalakrishnan (Footnote 94) 23.

[203] Ibid.

CHAPTER 3

Zero-Emissions Transport: Taxation Support for Business Electric Cars

Abstract Lowering of car transport emissions can facilitate enormous benefits for the local and global communities. This chapter considers the decarbonisation benefits of increasing zero-emissions vehicles in Australian business car fleets. It argues that more taxation support in Australia is necessary to achieve zero-emissions transport, with a focus on zero-emissions cars in business fleets. The Australian government has identified the light vehicle transport segment as one that needs to reduce its national CO_2 reductions. The chapter provides comparisons of European car tax concessions. Drawing on European success in policies that result in lower car emissions, the chapter demonstrates that Australian tax changes can lead to reduced CO_2 emissions and increase affordability for consumers, enabling a 'just transition' to decarbonisation.

Keywords Business electric vehicles · Just transition · Taxation law · Transport carbon emissions

D. Kraal et al., *A Just Transition to Decarbonisation*, Just Transitions,
https://doi.org/10.1007/978-3-031-49137-5_3

3.1 Introduction

High emissions from road transport is a recurring global theme raised at United Nations Climate Change Conferences, or Conferences of the Parties (COPs),[1] but the simple solution of transitioning to zero-emissions mobility is complex in practice. The price gap between fossil fuel and zero-emissions passenger transport vehicles is still too wide for many buyers. Enabling a shift to affordable, zero-emissions car transport via business fleet cars is a 'just transition' issue, as we need to alleviate past social inequalities in the move away from fossil fuels to renewable energy.

At COP26 in Glasgow in 2021, government and industry leaders debated and then committed to 100% zero-emissions new vehicle sales by 2040 (or earlier). This accelerated pace of transition is needed to meet the Paris Agreement goal of limiting global warming to 1.5–2 °C. In 2022, COP27 in Egypt had a similar theme embodied in its Decarbonisation Day. In late 2023, COP28 in Dubai continued the transport theme.[2]

This chapter's topic is taxation support in Australia for zero-emissions cars in business fleets—one of the identified transition pathways to decarbonisation.[3] The wider community can also contribute to lowering car emissions, as business fleet cars (including zero-emissions cars) are typically retired after three years and are much sought by the public as second-hand cars. A 'just transition' includes distributive justice, which addresses the allocation of benefits and negatives from sectors with high emissions.

Battery electric vehicles (BEVs) are a type of zero-emissions vehicle (ZEV). In Australia, almost half of all light vehicles are purchased by businesses, and with some tax support they could be ZEVs, but only if they are affordable. At COP26, 'ministers, vehicle manufacturers, business

[1] See 'transport' listed as a theme at *UN Climate Change Conference UK 2021*. https://ukcop26.org/the-conference/presidency-programme.

[2] UN Climate Change Conference UK 2021, *Presidency Programme: COP26*. https://webarchive.nationalarchives.gov.uk/ukgwa/20230311041857/https://ukcop26.org/the-conference/presidency-programme/. See also COP27 Presidency Vision, Thematic Days. https://cop27.eg/#/presidency/eventsThematic, and United Nations COP28 Themes. https://climatechampions.unfccc.int/category/cop28/.

[3] This concept has been applied by Amandine Denis-Ryan et al., *Pathways to Deep Decarbonisation in 2050: How Australia Can Prosper in a Low Carbon World* (ClimateWorks Australia, 2014). https://www.climateworkscentre.org/wp-content/uploads/2014/09/climateworks_pdd2050_initialreport_20140923-1.pdf.

leaders, fleet managers and city leaders signed a statement in agreement that this COP will kick-start the mass market for zero-emissions vehicles'.[4]

Cars are both a transport and energy product. Road transport contributed to 19% of Australia's CO_2 emissions in 2022, with passenger vehicles contributing almost 50% of those emissions. BEVs are a zero-emissions alternative; however, if their energy source is fossil fuel-generated electricity then it must be kept in mind that coal-fired electricity contributed to 32% of Australia's CO_2 emissions in 2020.[5] To balance concerns about CO_2 emissions from car charging, consider that energy stored in car batteries can also be used to power a home. Businesses need to transition to BEVs, which, if powered by renewable energy, are a better option than hybrid cars. It is sobering that the uptake of BEVs by business fleets in Australia was only 1.2% (around 400 in number) in 2020,[6] and electric vehicle (EV) sales represented only 3.8% of all new vehicle sales in 2022.[7]

In Australia, 40% of light passenger vehicles sold in 2020 were purchased by businesses.[8] Further, a survey showed that 47% of fleet vehicles are home garaged, providing an opportunity for business fleet BEVs to be home charged using smart meters to gain off-peak rates, avoid grid congestion and address the lack of business-site charging.[9] This points to the importance of enquiring whether taxation changes can facilitate affordable BEVs and charging infrastructure for business. Significantly,

[4] UN Climate Change Conference UK 2021, *Presidency Programme* (Footnote 2).

[5] Almost one-fifth of Australia's emissions are from transport; see Department of Industry, Science, Energy and Resources (DISER), Australian Government, *Quarterly Update of Australia's National Greenhouse Gas Inventory: September 2022.* Global emissions from the transport sector are around 23%; see IDDRI.org, 'Deep Decarbonization Pathways for Mobility'. www.iddri.org/en/project/deep-decarbonization-pathways-mobility.

[6] National Transport Commission (NTC), Australian Government, *Carbon Dioxide Emissions Intensity for New Australian Light Vehicles 2020* (NTC, 2021).

[7] Federal Chamber of Automotive Industries (FCAI), *Submission to the Australian Government on the Future Fuels Strategy Discussion Paper 2021.* https://consult.dcceew.gov.au/future-fuels-strategy.

[8] FCAI, *VFACTS December 2020: New Vehicle Sales Figures* (FCAI, 2020); DISER, *Future Fuels Strategy: Powering Choice* (Discussion Paper, Australian Government, 2021).

[9] Australasian Fleet Management Association (AFMA) and AGL Energy Ltd, *Survey: Electric Vehicles in Business Fleets* (AFMA and AGL, 2020). https://afma.org.au/.

BEVs in business fleets can eventually be shifted to the second-hand market and sold as affordable commodities to the wider community, enabling a 'just transition' decarbonisation.

A variety of affordable BEV models are needed in Australia, but poor emissions regulations have resulted in the country being a 'dumping ground' for cars with poor emissions controls.[10] One analysis notes that in the United Kingdom (UK) 29 BEV models cost less than AU60,000, but by 2021 there are only four such models in Australia.[11] With regard to BEV charging infrastructure, in the UK and the US, 30% of BEV charging events occur at the workplace.[12] Reiner's UK review concludes that participants are willing to pay more for a BEV if there is access to charging at work.[13]

Research data from Europe reveals that tax policies to encourage transitioning internal combustion engine vehicles (ICEVs) to equivalent BEVs can help reduce CO_2 emissions.[14] Australian research has compared equivalent BEVs and ICEVs to estimate the impact of switching to BEVs, indicating saved CO_2 emissions averaging 788 kg per annum. This figure came from 2022 data on annualised emissions from selected cars across Australia.[15] This suggests that more tax policies to support reducing

[10] Jake Evans, 'Australia Becoming a "Dumping Ground" for Polluting Cars as Government Delays Signing on to International Standards', *ABC News*, 14 October 2021. www.abc.net.au/news/2021-10-14/australia-dumping-ground-for-polluting-cars-euro-6-standards/100535418.

[11] Electric Vehicle Council, *Submission to the Australian Government on the Future Fuels Strategy Discussion Paper 2021.* https://consult.dcceew.gov.au/future-fuels-strategy.

[12] Debapriya Chakraborty, David S Bunch and Jae Hyun Lee, *What Factors Drive Commuters' Demand for Electric Vehicle Charging Infrastructure?* (Policy Brief, Institute of Transportation Studies, University of California, November 2020).

[13] Carolin Reiner et al., *Driving Forward the Electric Revolution: Considerations for Policy* (UK Government, 2020). https://assets.publishing.service.gov.uk/government/uploads/system/uploads/attachment_data/file/914111/driving-and-accelerating-the-adoption-of-electric-vehicles-in-the-uk.pdf.

[14] See Dataforce, *Transport and Environment: Company Car Report* (Dataforce, 2020) 60. https://www.dataforce.de/en/e-mobility-2021/?utm_referrer=https%3A%2F%2Fwww.google.com%2F.

[15] The author acknowledges collaborative research conducted in 2021 regarding necessary taxation and fiscal changes in Australia to support fleet electric vehicles; see Mortimore, Kraal et al., *Business Fleets and EVs: Taxation Changes to Support Home Charging from the Grid, and Affordability*, Fast-track Project (RACE for 2030 CRC, 2022) 107.

emissions are overdue. For instance, the states of Tasmania and South Australia, which have the greenest energy mix, report the most emissions reductions.[16]

A BEV is preferable to a hybrid vehicle, as the latter combines a battery (which can be powered by renewable energy) with fossil fuel for energy;[17] any increase in the uptake of BEVs would ideally be from renewable energy electricity. We need new technology to reduce CO_2 emissions but fiscal changes are argued to be a complementary strategy. Combining short- and long-term taxation and subsidy solutions will promote the uptake of BEVs to achieve the benefits of lower emissions.

The pathway of using taxation and subsidies aims at achieving a greater uptake of affordable fleet BEVs that are fit for purpose. Supporting home charging of business fleet BEVs at a fleet employee's place of residence is 'just'.

3.1.1 Decarbonisation's Contribution to a 'Just Transition'

The identification of both short- and long-term tax changes and other necessary fiscal subsidies will address the cost barrier arising from the price premium on BEVs and tax barriers inherent in establishing home charging for business fleet vehicles. Tax changes can lead to reduced CO_2 emissions and increase affordability for consumers, enabling a 'just transition' to a low-carbon economy.

Tax changes to support decarbonisation are quantified as follows. Currently a conservative 47% of fleet cars (around 35,000 cars) are home garaged.[18] It has been forecast that by 2030 there will be around 85,000 corporate fleet vehicles in Australia. If just 35,000 of those were BEVs,

[16] Tasmania predominantly uses hydro-electricity and South Australia prefers solar energy and large-scale batteries. Extreme rain events in eastern Australia in March 2022 forced a change in policy to net zero by 2030; see Paul Sakkal, '"We're Not the Party of 2018": State Liberals Adopt Net Zero Target', *The Age*, (8 March 2022). www.theage.com.au/politics/victoria/we-re-not-the-party-of-2018-state-liberals-adopt-net-zero-target-20220301-p5a0jn.html.

[17] Hydrogen fuel cell electric vehicles (FCEVs) are an alternative to battery electric vehicles (BEVs). An FCEV uses electricity from a fuel cell powered by compressed hydrogen.

[18] AFMA and AGL Energy Ltd (Footnote 9).

home charging would enable cost savings because of significant reductions in petrol usage; the estimated reduction in fossil fuel electricity emissions is 55,000 tonnes of CO_2 per year.[19]

3.1.2 *Chapter Roadmap*

This chapter covers relevant literature on BEV acquisition barriers that include charging infrastructure issues. This is followed by background information and discussion on high carbon emission sectors; policies in Australia to cover carbon emissions; and current tax support to increase the number of EVs in Australia. Then explored is current funding support for EVs and infrastructure, and information on EV sales to fleets. The section 'Lessons from Europe' is followed by the conclusion.

3.2 Literature: Battery Electric Vehicle Acquisition Barriers and Charging Infrastructure

The literature review that follows focusses on BEV acquisition barriers, and the merits of alternative BEV charging infrastructure sites—whether public, business or at home.

3.2.1 *Acquisition Barriers*

Acquisition barriers around the uptake of BEVs have been identified internationally, and Australia is in a similar situation. The reviewed literature covers consumer lack of trust in claims about the environmental benefits of BEVs, such as energy sources and the sustainability of batteries.[20] Other barriers include range anxiety stemming from a fear of battery

[19] By 2030, use of home-charged BEVs will result in estimated CO_2 emission reductions of 55,000 tCO_2/yr; data from the AFMA and AGL and cited in Mortimore, Kraal et al. (Footnote 15) 107.

[20] Mehmet Biresselioglu, Melike Kaplan and Barbara Yilmaz, 'Electric Mobility in Europe: A Comprehensive Review of Motivators and Barriers in Decision Making Processes' (2018) 109(C) *Transportation Research Part A: Policy and Practice* 1; *The Environmental Impact of Lithium Batteries*, Institute for Energy Research. www.instituteforenergyresearch.org/renewable/the-environmental-impact-of-lithium-batteries. See also Mortimore, Kraal et al. (Footnote 15) 143.

failure—a significant factor in consumer resistance to BEVs[21]—and supply issues resulting from manufacturers' low profitability on BEVs.[22] Studies blame these problems on a lack of consumer education about BEVs.[23] The inadequacy of the former Morrison federal government's policy on CO_2 emissions was commented on by submissions to the Australian government's *Future Fuels Strategy* inquiry in 2021.[24]

A barrier of particular significance in Australia is the price of BEVs. Indeed, the literature indicates that consumers give the most weight to purchase price.[25] Optimists claim that purchase price parity between BEVs and ICEVs will eventually occur with lower battery prices or at some politically acceptable juncture when government incentives are available to promote EVs.[26] While there are calls for governments to either subsidise BEVs or begin taxing ICEVs, the key point is that the significant purchase price difference between BEVs and ICEVs needs to be closed.[27] Broadbent et al. find that 'incentives positively influence the likelihood that private motor vehicle customers will buy an EV the next

[21] Peynan Ashkrof, Gonçalo Homem de Almeida Correia and Bart Van Arem, 'Analysis of the Effect of Charging Needs on Battery Electric Vehicle Drivers' Route Choice Behaviour: A Case Study in the Netherlands' (2020) 78 *Transportation Research Part D: Transport and Environment* 102,206.

[22] Electric Vehicle Council (Footnote 11); Biresselioglu et al. (Footnote 20); Zoe Long et al., 'What Does Tesla Mean to Car Buyers? Exploring the Role of Automotive Brand in Perceptions of Battery Electric Vehicles' (2019) 129(C) *Transportation Research Part A: Policy and Practice* 185.

[23] Gail H Broadbent, Graciela Metternicht and Danielle Drozdzewski, 'An Analysis of Consumer Incentives in Support of Electric Vehicle Uptake: An Australian Case Study' (2019) 10(1) *World Electric Vehicle Journal* 1; Xiang Zhang, Xue Bai and Jennifer Shang, 'Is Subsidized Electric Vehicles Adoption Sustainable: Consumers' Perceptions and Motivation toward Incentive Policies, Environmental Benefits, and Risks' (2018) 192 *Journal of Cleaner Production* 71; Lance Noel et al., 'Willingness to Pay for Electric Vehicles and Vehicle to Grid Applications: A Nordic Choice Experiment' (2019) 78(C) *Energy Economics* 525; Centre for New Energy Technologies, 'Electric Vehicle Update and Charging: A Consumer Perspective' (Webinar, 3 February 2021).

[24] See, e.g., Electric Vehicle Council *Submission* (Footnote 11); FCAI *Submission* (Footnote 7).

[25] Electric Vehicle Council (Footnote 11) covers capital and operating costs of a vehicle.

[26] Transport & Environment, *Hitting the EV Inflection Point*, 10 May 2021. https://www.transportenvironment.org/discover/hitting-the-ev-inflection-point/.

[27] Johannes Kester et al., 'Policy Mechanisms to Accelerate Electric Vehicle Adoption: A Qualitative Review from the Nordic Region' (2018) 94(C) *Renewable and Sustainable Energy Reviews* 719; Noel et al. (Footnote 23).

time they buy a car'.[28] Caution is needed nonetheless, for if demand for BEVs is solely driven by incentive policies, Zhang et al. claim that—as seen in Denmark and the US state of Georgia—large-scale BEV adoption does not continue after subsidies phase out.[29]

A key issue to the uptake of BEVs is charging infrastructure availability, whether at public facilities, at business sites or at home. This is covered in the next section.

3.2.2 Charging Infrastructure

3.2.2.1 Public Charging

Studies have noted the importance of public fast-charging opportunities to increase adoption of BEVs.[30] In response, in 2021 the Australian government offered funding for public charging infrastructure.[31] Nonetheless, Ashkrof et al. claim that for many BEV drivers, charging at home or at business sites may be preferable over public charging, because of the convenience and lower cost.[32]

Public charging infrastructure deployment might grow enough to entice consumers towards a BEV, but 'free parking' public charging, such as at supermarkets, has been criticised as not sustainable[33] and, as noted by Lavieri and Bas Domenech, there is the problem of charge point congestion or 'hogging' of free chargers.[34]

[28] Broadbent et al. (Footnote 23) 4.

[29] Zhang et al. (Footnote 23).

[30] Ashkrof et al. (Footnote 21) 2; Broadbent et al. (Footnote 23) 11.

[31] DISER, *Future Fuels and Vehicles Strategy: Powering Choice*, Final Report (Australian Government, 2021) 1.

[32] Ashkrof et al. (Footnote 21).

[33] Gerardo Zarazua de Rubens et al., 'The Market Case for Electric Mobility: Investigating Electric Vehicle Business Models for Mass Adoption' (2019) 194 *Energy* 116841.

[34] Patricia Lavieri and Carmen Bas Domenech, *Electric Vehicle Uptake and Charging: A Consumer-Focused Review* (University of Melbourne, 2021).

Public charging needs to be provided in dense urban areas. It is also needed where workplace charging is restrictive, or for fleets such as taxis and ride-sharing services (e.g. large charging hubs, such as those already existing in China).[35]

3.2.2.2 *Business-Site Charging*

Companies may be willing to install charging infrastructure in their parking facilities as part of corporate social responsibility policies, staff and visitor BEV schemes, or even BEV governance requirements of companies.[36] However, according to Chakraborty et al., in the UK and US only 30% of BEV charging events occur at the workplace.[37]

Reiner's review concludes that car drivers are willing to pay more for a BEV if there is access to workplace charging.[38] Walz et al. examine the use of workplace BEV charging and its effects on grid planning, observing that the trade-off between convenience and monetary savings translated into substituting home charging with free workplace charging. Walz et al. identify several business-site variables affecting workplace charging, such as working time, distance to work, home charging possibility and charging power, and car model. A higher number of charging stations available at the workplace can make this option more attractive.[39] The Australian government has noted that, nationally, there are few business-site charging infrastructure facilities.[40]

3.2.2.3 *Home Charging*

Debate on the option of charging BEVs at home indicates the importance of developing the design, tools and policies to resolve barriers related to

[35] International Energy Agency (IEA), *Global EV Outlook 2020: Entering the Decade of Electric Drive?* (IEA, 2020).

[36] Tianjin Chen et al., 'A Review on Electric Vehicle Charging Infrastructure Development in the UK' (2020) 8(2) *Journal of Modern Power Systems and Clean Energy* 193.

[37] Chakraborty et al. (Footnote 12).

[38] Reiner et al. (Footnote 13).

[39] Kathrin Walz, Daniel Contreras and Krzysztof Rudion, 'Modelling of Workplace Electric Vehicle Charging Profiles Based on Trip Chain Generation' (Conference Paper, IEEE PES Innovative Smart Grid Technologies Europe, 26–28 October 2020); see also Lavieri and Bas Domenech (Footnote 34).

[40] DISER, *Future Fuels Strategy*, Discussion Paper (Footnote 8).

charging at residences.[41] In Australia, Lavieri and Bas Domenech find that home charging (or overnight charging near home when home charging is not available) is the preferred option. The second most popular charging location is the workplace, followed by commute-related charging points (e.g. public transport hubs, park-and-ride facilities).[42]

A US survey shows that 72% of potential EV users consider charging a car overnight twice per week to be highly convenient; while only 36% of potential users consider charging at *fast* Level 3 (wait time of 30 minutes) public charging stations once a week to be very convenient.[43] A US report by Bui et al. finds that most EV charging is likely to continue at home, where it is less expensive and more convenient than public options.[44] A government of Ireland report states that home charging is considered the primary method of charging for the majority of EVs and is considered least expensive. It accounts for around 80% of EV charging sessions in Ireland and will continue to be the primary method of charging in the future.[45]

Reiner et al. conclude for the UK that it is most important to have EV charging infrastructure at home, followed by the workplace and then public locations.[46] The Klein et al. Europe-wide report finds that the perceived benefit of EV home charging decreases continuously with faster charging times at public charging stations.[47]

In the case of Canada, Lavieri and Bas Domenech note that EV users are likely to overlook residential charging for free destination charging

[41] Chen et al. (Footnote 36).

[42] Lavieri and Bas Domenech (Footnote 34).

[43] Union of Concerned Scientists (UCSUSA) and Consumer Reports, *Electric Vehicle Survey Findings and Methodology* (UCSUSA, 2019). www.ucsusa.org/sites/default/files/attach/2019/07/2019-EV-Survey.pdf.

[44] Anh Bui, Peter Slowik and Nic Lutsey, *Los Angeles Electric Vehicle Charging Infrastructure Needs and Implications for Zero-Emission Area Planning* (White Paper, International Council on Clean Transportation, 2021).

[45] Government of Ireland, *Transport Energy*. https://www.gov.ie/en/policy-information/e1539-transport-energy/.

[46] Reiner et al. (Footnote 13).

[47] Marvin Klein, Christine Strauss and Christian Stummer, 'Business Information Through Choice-Based Conjoint Analysis: The Case of Electric Vehicle Home Charging', in N Kryvinska and A Poniszewska-Marańda (eds), *Developments in Information & Knowledge Management for Business Applications: Volume 2* (Springer, 2021) 357.

when looking for monetary savings.[48] They find that owners of BEVs with driving ranges greater than 320 km, such as Tesla models, are more likely to use only home charging. Individuals who have solar panels at home are also more likely to rely exclusively on home charging given the potential for electricity cost savings.[49]

3.2.2.4 *Availability*

Research on car charging potential for the Australian state of Victoria finds that up to 78% of households would prefer charging at home.[50] In contrast, in the UK, Grote et al. find that access to charging infrastructure in residences and on-street parking areas is a barrier to widespread BEV uptake.[51] Even where large areas of urban on-street parking exist, a close-to-home parking space may not be routinely available and, when it is, charging cables crossing a public footpath from property to vehicle is a safety and security risk.

3.2.2.5 *Charging Power*

In Europe, BEVs can be charged using household power points through *standard* Level 1 charging. In Norway, 63% of surveyed BEV drivers use Level 1 charging at home.[52] Chen et al. claim that, although 35% of UK homeowners charge at home, they only have Level 1 power.[53] Level 2 public chargers cost up to 15 times more than Level 2 at-home chargers.[54]

According to researchers using US data, even though some BEV users only have Level 1 charging in their homes, the penetration of Level 2

[48] Lavieri and Bas Domenech (Footnote 34) 32.

[49] Ibid.

[50] Centre for New Energy Technologies (Footnote 23).

[51] Matt Grote et al., 'Locating Residential On-street Electric Charging Infrastructure: A Practical Methodology' (2019) 74(3) *Transportation Research Part D: Transport and Environment* 15.

[52] Australian Energy Market Commission, *2020 Retail Energy Competition Review*, Final Report (30 June 2020) 199.

[53] Chen et al. (Footnote 36).

[54] Chakraborty et al. (Footnote 12).

charging is rapidly increasing as this seems to be users' preferred residential option.[55] In Australia, a national government energy body expects BEV battery prices to reduce and battery capacity to increase.[56] As faster charging infrastructure is preferred, for Australia it is likely there will be more Level 2 charging at home and then a switch to more powerful Level 3 public chargers. Chargers will have future generation software to enhance communication between vehicle and charger, as well as across charging networks.

3.2.3 *Literature Summary*

The high purchase price of BEVs and range anxiety in Australia are barriers of particular significance. The literature indicates that consumers give more weight to the upfront purchase price compared to other acquisition barriers. Regarding BEV charging infrastructure, in the short term there is a preference for home charging over public or business-site facilities. Home charging is preferred in the UK, Ireland, US and Australia because of its higher convenience and lower cost compared with public options. Overall, the literature provides a starting point for the consideration of taxation for delivering a greater uptake of affordable fleet BEVs.

3.3 Background to Australia's Support for Electric Vehicles

In February 2021 Australia's (former) Morrison government released its *Future Fuels Strategy* discussion paper to the public, setting out practical actions to enable businesses to deploy low-emissions road transport.[57] The paper promotes a mix of fuels and vehicle technologies as the path to the future. Practical actions focus on light passenger vehicles (under 1 tonne) in fleets, as they represent 40% of new vehicles on the road. The paper covers EVs, but notes barriers—such as large capital costs

[55] Lavieri and Bas Domenech (Footnote 34) 34. Level 2 charging adds about 25 miles of range per hour. Level 1 requires almost a full day to charge an EV, which is just too slow for most drivers.

[56] Australian Energy Market Commission (Footnote 52).

[57] DISER, *Future Fuels Strategy*, Discussion Paper (Footnote 8).

and a lack of charging infrastructure—that cannot be addressed by technology alone. The government stated that where possible, it would look to the Australian Taxation Office (ATO), as the country's taxation system regulator, to provide guidance to 'normalise electric vehicle expenses to internal combustion vehicle expenses in public advice'.[58] The discussion paper does not raise the issue of legislating new tax changes to support EVs.

In November 2021, the former Morrison government's final report, *Future Fuels and Vehicles Strategy*, was released.[59] It confirms the 'fleets first' approach, but retains an emphasis on 'technology not taxes' to reduce road transport CO_2 emissions.[60] The report implies that taxpayers must rely on existing taxation provisions, including 'temporary full expensing'; 'instant asset write-off' (for both, 100% of an asset's capital cost can be deducted in one financial year, over limited timeframes); and accelerated depreciation. It references a 2021 ATO publication for guidance on how to leverage existing taxation provisions to reduce energy costs.[61]

Australia elected a new 'climate-friendly' federal government in May 2022, headed by Prime Minister Anthony Albanese. On 19 October 2022, the new government announced plans to fast-track multi-billion-dollar renewable energy projects in the states of Victoria and Tasmania.[62] Days later, the Albanese government's October Budget 2022–2023 included total climate-related spending of AU$25 billion from 2022–2023 to 2029–2030.[63]

[58] Ibid., at 19.

[59] DISER, *Future Fuels Strategy*, Final Report (Footnote 31).

[60] Ibid., at 6, 16.

[61] Energy Efficiency Council (EEC), *Leveraging Tax Incentives to Improve Energy Performance. A Guide for Australian Businesses Investing in Smart Energy Management* (EEC, 2021) 179.

[62] Phil Mercer, 'Australia Pledges Multi-Billion Dollar Renewable Energy Projects', *Voice of America East Asia*, 19 October 2022. https://www.voanews.com/a/australia-pledges-multi-billion-dollar-renewable-energy-projects-/6796061.html.

[63] Commonwealth Government, Budget: October 2022–23—Budget Strategy and Outlook (Budget Paper No 1, 25 October 2022) 15–19. https://budget.gov.au/2022-23-october/content/bp1/index.htm.

To encourage the purchase of EVs, the Albanese government proposed a parliamentary bill (passed later in 2022) exempting EVs from the provisions in the *Fringe Benefits Tax Assessment Act 1986* (FBT). For instance, the purchase of an EV valued at AU$50,000 and held for three years (as is typical), could save an employer up to AU$9,000 a year in FBT. In Australia, the value of fringe benefits provided to employees is subject to tax, but this is paid by the employer. The most popular fringe benefits provided to employees are light passenger vehicles. The FBT rate of 47% is based on the highest marginal tax rate plus the Medicare levy. The next section discusses EV policy in the states, territories and federal governments in Australia.

3.4 Policy Discussion

3.4.1 High Carbon Emission Sectors

In 2022, the Australian transport sector's share of CO_2 emissions was 19% of national emissions, around 50% of which was from the 20 million registered cars on the road. Over the same timeframe, coal-fired electricity generated the highest CO_2 emissions, at 32% of national emissions.[64]

Australia's transport emissions are slightly lower than those from the European Union transport sector, which in 2019 contributed almost 23% of Europe's total CO_2 emissions.[65] The key difference is that in 2019, Europe had 313 million registered cars on the road,[66] compared with 20 million in Australia. The European Environment Agency has reported that the gas and coal-fired electricity sector in Europe generates the highest CO_2 emissions: around 26% of total European emissions in 2019.[67] A better measure of light vehicle transport emissions is grammes

[64] Department of Climate Change Energy the Environment and Water, 'Quarterly Update of Australia's National Greenhouse Gas Inventory: Sept 2022' (Australian Government, 2022).

[65] European Environment Agency, EEA Greenhouse Gases Viewer, 2021. https://www.eea.europa.eu/data-and-maps/data/data-viewers/greenhouse-gases-viewer.

[66] European Automobile Manufacturers Association (ACEA), 'Vehicles in Use in Europe 2022' (ACEA, 2022). https://www.acea.auto/files/ACEA-report-vehicles-in-use-europe-2022.pdf.

[67] European Environment Agency (Footnote 65) 185.

of CO_2 emitted per kilometre (gCO_2/km). In 2019 in Australia, light vehicle transport emissions for light vehicles averaged 181 gCO_2/km; in Europe it was a lower 122.3 gCO_2/km.[68]

3.4.2 *Current Tax Support to Increase the Number of Electric Vehicles*

3.4.2.1 *States' and Territories' Support*

The states and territories of Australia provide some tax policies that directly support the increase in absolute numbers of EVs. For instance, New South Wales (NSW) supports the purchase of all types of EV through stamp duty exemptions for new cars under AU$78,000 purchased after September 2021.[69] Australian states and territories do not subsidise EV home charging equipment,[70] although some have rebates for solar panels. The states' and territories' policies and legislation change often, and are not uniform. What is needed is a national approach, as covered in the next section.

3.4.2.2 *Federal Support*

Under the (former) Morrison government, the only support for EVs was through technology developments for infrastructure, not taxation

[68] European Environment Agency, CO_2 Performance of New Passenger Cars in Europe, 2021. https://www.eea.europa.eu/ims/co2-performance-of-new-passenger; NTC (Footnote 6).

[69] Government of NSW, NSW Budget 2021–22: Revenue, 2021. https://www.budget.nsw.gov.au/sites/default/files/2021-06/4.%20Review-BP1%20Budget%202021-22.pdf; Revenue NSW, *Grants and Schemes—Electric Vehicles*, 2022. https://www.revenue.nsw.gov.au/grants-schemes#electric-vehicles.

[70] Eg. Australian Capital Territory, Queensland and Northern Territory Governments: ACT Government (2022) Zero Emissions Vehicles. https://www.environment.act.gov.au/cc/zero-emissions-vehicles; Mick de Brenni, Media release 21 April 2023. 'Making electric vehicles more affordable to slash emissions and act on climate change'. https://statements.qld.gov.au/statements/97613; Queensland Government (2022) 'Shifting to Zero Emission Vehicles'. https://www.qld.gov.au/transport/projects/electricvehicles/hitting-the-road; and Northern Territory Government (2022) 'Electric vehicle registration'. https://nt.gov.au/driving/rego/fees/registration-fees#electric_vehicle.

policy.[71] Since the May 2022 election of the Albanese government, however, federal taxation support for EVs has improved.

The change began in July 2022 when the *Treasury Laws Amendment (Electric Car Discount) Bill 2022* (the Bill) was introduced to the Federal House of Representatives for the first time.[72] The Bill applies an FBT exemption to fleet passenger cars (under 1 tonne). Fleet vehicles typically transition as cheaper cars to the second-hand market after around three years.

In November 2022, the Bill was passed by both the House and the Senate, and upon royal assent, the Act took effect from July 2022.[73] These changes will make EVs more affordable but more could be done.

In April 2023 the Albanese government announced its *National Electric Vehicle Strategy*, designed to increase the uptake of EVs and to reduce emissions.[74] The strategy aims to increase the supply of affordable EVs; establish more EV charging infrastructure[75]; and encourage increased EV demand by introducing fuel efficiency standards.

The supporting strategy document claims that implementing these initiatives will help expand EV availability and choice; reduce road transport emissions; make it easier to charge EVs across Australia; and make EVs more affordable.[76] Fuel efficiency standards (i.e. emissions standards) will be a key element, to which industry input was invited during 2023.

[71] Scott Morrison, 'Prime Minister's Address at the National Press Club', 1 February 2021, National Press Club, Canberra. https://www.pm.gov.au/media/address-national-press-club-barton-act.

[72] Parliament of Australia, *Treasury Laws Amendment (Electric Car Discount) Bill 2022*. Treasury Laws Amendment (Electric Car Discount) Bill 2022—Parliament of Australia (aph.gov.au).

[73] Treasury Laws Amendment (Electric Car Discount) Act 2022.

[74] Chris Bowen and Catherine King, 'Australia's first National Electric Vehicle Strategy to Drive Cleaner, Cheaper-to-Run Vehicles' (Joint Media Release, 19 April 2023). https://minister.dcceew.gov.au/bowen/media-releases/joint-media-release-australias-first-national-electric-vehicle-strategy-drive-cleaner-cheaper-run-vehicles.

[75] Chris Bowen, 'Delivering More EV Charging to Make Cleaner, Cheaper Cars More Accessible' (Media release, 19 April 2023). https://minister.dcceew.gov.au/bowen/media-releases/delivering-more-ev-charging-make-cleaner-cheaper-cars-more-accessible.

[76] Department of Climate Change Energy the Environment and Water, *The National Electric Vehicle Strategy* (Australian Government, 2023).

3.4.3 *Current Funding Support for Electric Vehicles and Infrastructure*

The Australian government's *Budget 2022–23*, handed down in October 2022, includes total climate-related spending of AU$25 billion from 2022–2023 to 2029–2030.[77] The Albanese government's budget is part of the Paris Agreement commitment efforts to lower Australia's CO_2 emissions by 43% below 2005 levels by 2030. The budget's major climate-related spending is earmarked for a range of projects, of which eight can be classified as demand-side projects. These include the *Driving the Nation Fund* (AU$275.4 million) designed to deliver EV public charging infrastructure; the FBT exemption for EVs' and 5% import tariff exemptions.

3.4.4 *Sales of Electric Vehicles to Fleets*

Fleets in Australia represented 40% of new car acquisitions in 2020, yet fleet acquisitions of EVs numbered a mere 487. The wider community's purchase of all types of EVs was still low, at 3,470 cars in 2020; and fleet acquisitions of BEVs in Australia in 2020 were a low 1.4% of total EV sales.[78]

The preceding discussions on Australia's emissions statistics and its federal, state and territory tax policies are summarised in Table 3.1. The figure also provides some comparative data from selected European countries (Norway, Germany, the UK and the Netherlands), which are explored in the next section.

3.5 Lessons from Europe

The following discussion takes account of EV tax policies for selected European countries (Norway, Germany, the Netherlands and the UK) as shown in Table 3.1.

The higher number of EVs in fleets in Europe than in Australia provides the latter with an aspirational target. European fleet data show that, for 21 countries, over 553,000 new light passenger EVs were

[77] Australian Government, *Budget 2022–23* (25 October 2022) 15–19. https://budget.gov.au/2022-23-october/content/bp1/index.htm.

[78] NTC 2021 (Footnote 6) 25, 57.

Table 3.1 Australia and Europe (Norway, Germany, UK and the Netherlands): selected comparative data on Tax and EVs

Item	*Description*	*Australia*	*Europe*
1	Share of CO_2 emissions from the transport sector	19.0% (to September 2022)	23.0% (2019)
2	Share of CO_2 emissions from the electricity sector	32.0% (to September 2022)	26.2% (2019)
3	Current tax support to increase the number of EVs: state and federal	– NSW, Vic, ACT, Qld and NT have tax policies/subsidies to directly support the purchase of EVs – SA, WA and Tasmania have fiscal policies to support charging infrastructure – The former Morrison federal government supported EV infrastructure technology only – The Albanese federal government legislated FBT exemption for fleet cars from July 2022 – The Albanese *National Electric Vehicle Strategy* 2023 might have tax penalties as part of its fuel emissions standards	– The selected European countries link tax penalties to high-emission vehicles – The EU regulates new car performance standards to achieve low emissions
4	Government policies with a high impact on EV uptake	– NSW policies have the most positive impact on EV uptake – NSW made up almost 32% of national EV sales in 2019 and 2020	– Norway's policies have the most positive impact on EV uptake in Europe, followed by the UK – Norway will ban sales of new ICEVs from 2025 and the UK, from 2030

Item	*Description*	*Australia*	*Europe*
5	Sales statistics for EVs: fleets	487 new EVs for fleets, nationally (2020)	553,000 new EVs for fleets in Europe (2020); this was 54% of all new EVs in Europe
6	Sales statistics for EVs: all buyers	3,470 new EVs for all buyers, nationally (2020)	1,024,074 new EVs for all buyers in Europe (2020)

Source Authors. Note abbreviations: Victoria (Vic); Australian Capital Territory (ACT); Queensland (Qld); Northern Territory (NT); South Australia (SA); Western Australia (WA)

registered by companies in 2020. This was 54% of all EV car registrations.[79] The remaining 46% of EVs (471,000) in 2020 were bought by private individuals. This suggests that European tax policies, particularly those in the selected European countries, have been more successful than Australian tax policies at lowering CO_2 emissions.

Figure 3.1 shows that the selected European countries in 2019 reported emissions from passenger vehicles of around 122 gCO_2/km, in contrast to Australia generating a very high 181 gCO_2/km.[80] This provides evidence of those selected European countries successfully lowering CO_2 emissions in light vehicle transport via tax policies.

Electric cars accounted for 17% of Europe's auto sales in 2020. According to the International Energy Agency's *Global EV Outlook 2020*

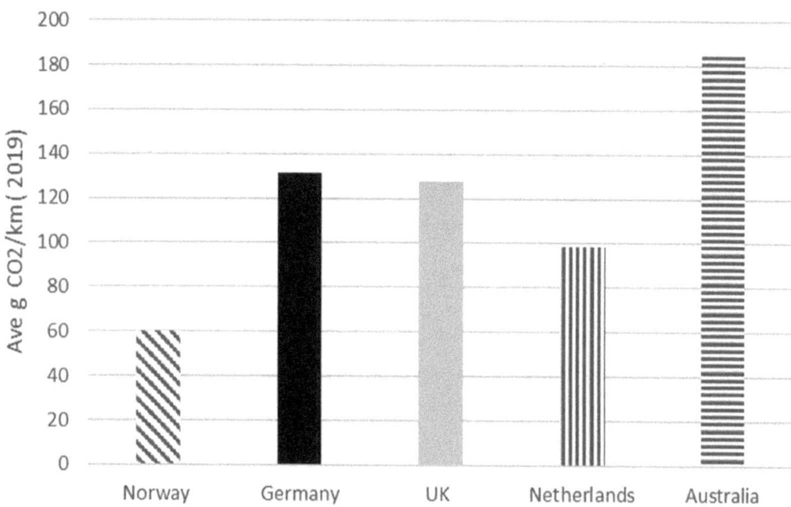

Fig. 3.1 Carbon emissions for passenger vehicles: selected countries, 2019 (*Source* Authors)

[79] ACEA, 'Vehicles in Use in Europe 2022' (Footnote 66).

[80] ACEA, *The Automobile Industry Pocket Guide 2020–2021* (ACEA, 2020). For Australian data, see NTC, 2021 (Footnote 6).

report, 'the key driver underpinning EV growth in Europe is the tightening CO_2 emissions standards that occurred in 2020'.[81] The report also notes that 'the expansion of purchase subsidies and tax benefits in major markets also contributed to the acceleration of sales'.[82] The selected European jurisdictions commonly provide stamp duty and annual registration concessions to BEVs, with the UK also providing capital allowance and congestion tax concessions for BEVs.[83] The selected countries link tax penalties to high-emission vehicles in the levying of road user taxes and 'private use' of employer-provided cars. Europe maintains ambitious emission performance standards for new cars.[84]

3.5.1 Norway

Norway's tax policies have the most positive impact in Europe on EV uptake.[85] Car registration tax exemption and value-added tax (VAT) concessions have been particularly successful in Norway.[86] In 2020, Norwegian company car fleets had around 25,000 EVs, split between PHEVs (plug-in hybrid electric vehicles) and BEVs.[87] Norway will ban sales of new ICEVs from 2025.[88]

[81] IEA, *Global EV Outlook 2020* (Footnote 35) 17.

[82] Ibid.

[83] PriceWaterhouseCoopers (PWC), *Global Automotive Tax Guide 2020* (PWC, 2020); European Automobile Manufacturers Association (ACEA), Overview—Electric vehicles: Tax benefits & purchase incentives in the European Union (ACEA, 2022) https://www.acea.be/publications/article/overview-of-incentives-for-buying-electric-vehicles, 77.

[84] European Commission (EC), 'CO_2 Emission Performance Standards for Cars and Vans' (EC, 2021). https://climate.ec.europa.eu/eu-action/european-green-deal/delivering-european-green-deal/co2-emission-performance-standards-cars-and-vans_en#.

[85] ACEA, 'Vehicles in Use in Europe 2022' (Footnote 66) 14.

[86] Norwegian Government, 'Norway is Electric' (Ministry of Transport, 2021). https://www.regjeringen.no/en/topics/transport-and-communications/veg/faktaartikler-vei-og-ts/norway-is-electric/id2677481/; VATGlobal, 'Norway VAT Exemption Drives Electric Vehicle Boom' *VAT IT* (Blog Post, 2021). https://blog.vatit.com/blog/norway-drives-change-to-vat-on-electric-cars.

[87] DataForce, 'Company Car Report' (2020). https://www.dataforce.de/en/e-mobility-2021/?utm_referrer=https%3A%2F%2Fwww.google.com%2F; S. Wappelhorst, 'Small but Mighty: The Netherlands Leading Role in Electric Vehicle Adoption', *The International Council on Clean Transportation* (Blog Post, 2021). https://theicct.org/blog/staff/netherlands-ev-leader-feb2021.

[88] Elbil.no, *Norwegian EV Policy*. https://elbil.no/english/norwegian-ev-policy/.

Norway leads European EV sales at almost 65% of its automotive market in 2021.[89] The country relies on tax cuts and investment in publicly administered EV charging infrastructure, rather than direct subsidies. Norway exempted EVs from purchase tax and VAT until 2023; now these are concessional. Norway exempted EVs from road tax between 1996 and 2021.[90] Company car tax on EVs is lower than that on ICEVs. At the local level, counties and municipalities must reduce road and ferry tolls and parking charges by at least 50% for EVs.[91]

3.5.2 Germany

Germany is now delaying its ban on sales of new ICEVs to 2035, as support for this policy is declining. Nonetheless, Germany offers some of the most generous EV incentives in Europe, with a goal of having 10 million EVs and 1 million charging stations by 2030.[92]

German incentives include a €6,000 grant for purchase of an EV priced up to €40,000 (the grant is reduced to €5,000 for vehicles priced between €40,000 and €65,000).[93] Those incentives are scheduled to reduce to €4,500 and €3,000, respectively, in 2023 because the German government believes that EVs' growing popularity makes the subsidies

[89] Mathilde Carlier, 'Share of Fuel Types of Passenger Cars in Europe by country 2022', *Statistica.com*. https://www.statista.com/statistics/1128792/passenger-car-registrations-in-europe-by-fuel-type/; EEA, 'New Registrations of Electric Vehicles in Europe'. https://www.eea.europa.eu/ims/new-registrations-of-electric-vehicles.

[90] Ibid.

[91] Ibid.

[92] Walbox.com, 'EV and EV Charger Incentives in Europe' (Blog Post). https://blog.wallbox.com/en/ev-and-ev-charger-incentives-in-europe-a-complete-guide-for-businesses-and-individuals/.

[93] Germany Federal Ministry for Economic Affairs and Climate Action, 'Electric Mobility: Regulatory Environment and Incentives for Using Electric Vehicles and Developing a Charging Infrastructure'. https://www.bmwk.de/Redaktion/EN/Artikel/Industry/regulatory-environment-and-incentives-for-using-electric-vehicles.html; Reuters, 'Germany to Reduce Electric Car Subsidies in 2023', 26 July 2022. https://www.reuters.com/technology/german-coalition-parties-agree-reduce-e-car-subsidies-handelsblatt-2022-07-26/.

less necessary.[94] EVs registered between 2011 and 2030 receive a 10-year exemption from motor vehicle tax.[95] The company car tax is also discounted for EVs at 0.25% of the list price per month, compared with 1% of the list price for an equivalent ICEV.[96] There are also local purchase grants and charging infrastructure incentives, as well as free parking in some communities.[97]

3.5.3 *The Netherlands*

The Netherlands has the largest number of EV charging points in the EU.[98] As a consequence, range anxiety is a minor issue among the country's EV drivers.[99] Due to the Netherlands' incentive structure—which is primarily focussed on tax provisions—the cost of EV ownership has reduced, and now is similar to that of ICEV ownership.[100] One 2022 study predicts that Norway and the Netherlands will have close to 100% EVs in new light vehicle registrations by 2035.[101] EVs are exempt from purchase tax[102] and ownership tax.[103] ICEV owners pay additional

[94] Ibid.

[95] Allianz-Autowelt, 'Road Tax From 2023: Which Vehicles are Exempt From Road Tax or Tax-deductible?'. https://www.allianz-autowelt.de/kfz-steuer/rechner/#welche-fahrzeuge-sind-steuerbefreit.

[96] Umschalten, 'E-car Tax: You Should Know About These Tax Advantages for Your Company Car in 2022!'. https://www.umschalten.de/en/e-auto-steuer-vorteile-2022.

[97] Germany Federal Ministry for Economic Affairs and Climate Action, 'Electric Mobility: Regulatory environment and incentives for using electric vehicles and developing a charging infrastructure'. https://www.bmwk.de/Redaktion/EN/Artikel/Industry/regulatory-environment-and-incentives-for-using-electric-vehicles.html.

[98] Jose Pontes, 'The Netherlands—Where EV Charging Points Grow Like Tulips' 23 November 2021. https://autovista24.autovistagroup.com/news/the-netherlands-where-charging-stations-grow-like-tulips/.

[99] Ibid.

[100] Ibid.

[101] Electrek, 'New Study Predicts Which European Countries Will Have the Highest Percentage of New EVs by 2035'. https://electrek.co/2022/01/24/new-study-predicts-which-european-countries-will-have-the-highest-percentage-of-new-evs-by-2035/.

[102] Netherlands Electrics, 'BPM'. https://nederlandelektrisch.nl/subsidies-financiering/bpm.

[103] Netherlands Electrics. https://nederlandelektrisch.nl/subsidies-financiering/motorrijtuigenbelasting-mrb.

company car tax.[104] Moreover, the Netherlands imposes increased taxes on vehicles that emit high levels of CO_2.[105] Older diesel vehicles must pay a particulate matter surcharge.[106] After 2030, there will be a ban on new sales of ICEVs and only ZEVs may be newly registered in the Netherlands.[107]

3.5.4 *United Kingdom*

The UK's national laws mirror EU regulations on greenhouse gas emissions.[108] The UK has proposed an electrification strategy called the *Road to Zero* strategy.[109] Prime Minister Rishi Sunak aims to ban sales of new ICEVs from 2035.[110] Some of the UK's EV tax benefits include exemption of ZEVs from vehicle excise tax; full deduction against corporate tax for the cost of company cars that emit less than 50 gCO_2/km; and reductions in company car tax for EVs. EV purchasers can receive grants up to a maximum of £2,500 for vehicles priced under £35,000.[111] Some localities provide free parking for EVs and EVs are exempt from London's congestion charge until 2024.[112]

[104] Netherlands Electrics, 'Additional tax on company cars in 2023'. https://nederlandelektrisch.nl/subsidies-financiering/bijtelling.

[105] Netherlands Government Enterprise Agency, 'Motor Vehicle Tax'. https://business.gov.nl/regulation/motor-vehicle-tax/.

[106] Ibid.

[107] Netherlands Government Enterprise Agency, 'Electric Transport in the Netherlands'. https://english.rvo.nl/information/electric-transport.

[108] Hannah Cooper, 'Leaving the European Union: UK Climate Change Policy' (Library Briefing, House of Lords). https://researchbriefings.files.parliament.uk/documents/LLN-2017-0022/LLN-2017-0022.pdf.

[109] UK Department for Transport, 'The Road to Zero', 2018. https://assets.publishing.service.gov.uk/government/uploads/system/uploads/attachment_data/file/739460/road-to-zero.pdf.

[110] UK Government, 'Transitioning to Zero Emission Cars and Vans: 2035 Delivery Plan'. https://www.gov.uk/government/publications/transitioning-to-zero-emission-cars-and-vans-2035-delivery-plan.

[111] Ibid.

[112] Royal Automobile Club, UK, 'Do Electric Cars Pay the Congestion Charge?', 29 March 2022. https://www.rac.co.uk/drive/electric-cars/running/do-electric-cars-pay-the-congestion-charge/.

Table 3.2 draws from the preceding discussion across the selected EU countries to summarise European fiscal support, including regulations on new car performance standards that are linked to tax laws. It shows the laws in column one, with the coloured cells indicating countries with those laws in place. These European tax concessions provide a guide to frame further tax changes for Australia.

Table 3.2 European tax concessions, subsidies and new car performance policies

Item	*Norway*	*Germany*	*UK*	*Netherlands*
EV purchase subsidies: individual				
EV purchase subsidies: business				
EV VAT concession				
EV stamp duty on vehicle, exemption				
EV annual vehicle registration charge, exemption				
EV capital allowance concessions				
EV congestion tax, exemption				
Road tax based on CO_2 emissions				
Company car tax based on CO_2 emissions				
Ban on new gas/diesel cars	Yes, 2025	Perhaps, 2035	Yes, 2035	Yes, 2030

Source Authors

3.6 Conclusion

This chapter has been concerned with taxation support in Australia that is necessary to achieve zero-emissions transport, with a focus on increased zero-emissions cars in business fleets—one of the identified transition pathways to decarbonisation.

In 2021, the COP26 member governments and industry leaders committed to 100% zero-emissions new vehicle sales by 2040 (or earlier). This commitment was carried forward to COP28 in Dubai in 2023. The passenger car transport sector is a central plank of a 'just transition' to decarbonisation.

Regarding the EU countries selected for comparison, Norway's VAT concessions have been particularly successful, as have its vehicle purchase subsidies. European tax concessions for BEV stamp duty and annual registration concessions are common. The UK also provides capital allowance and congestion tax concessions for BEVs. The selected EU countries link tax penalties to high-emission vehicles for the levying of road user taxes and 'private use' of employer-provided cars. The EU maintains ambitious emission performance standards for new cars.

The Australian government's *National Electric Vehicle Strategy 2023* is designed to increase the uptake of EVs and introduce emissions standards. It will be a test of seeing policy through to legislation. The government must persist with its tax legislation programme to support the uptake of EVs. BEV business fleet cars can drive a 'just transition' for the benefit of the wider community via the second-hand car market. For instance, there are no plans yet to ban sales of new ICEVs through taxation measures. Decarbonisation of the transport system requires government tools such as taxation to incentivise the transition to ZEVs to achieve a low-carbon economy.

CHAPTER 4

Water Is Life: Recognising First Nations in Sustainable Water Management and Use

Abstract Taking its cue from the COP special theme of nature, this chapter highlights the importance of sustainable water management and use, and the vital role of First Nations people in that space in tackling climate change through decarbonisation. Through a just transitions lens, this chapter will examine the current water regulatory and policy framework in Australia, due to Australia's status as the driest inhabited continent on earth and in recognition that Australia's First Nations are acknowledged as having the oldest living culture on earth. In doing so it will consider the scope it provides for First Nations participation in water management and the recognition of First Nations cultural knowledge and use rights. It will conclude with some observations about the adequacy of Australian law and policy in that respect, and potential lessons for other nations, with a view to facilitating a more sustainable, just and equitable transition to a decarbonised world for First Nations peoples.

Keywords Indigenous peoples · First Nations people · Water management · Just transition

D. Kraal et al., *A Just Transition to Decarbonisation*, Just Transitions, https://doi.org/10.1007/978-3-031-49137-5_4

4.1 Introduction

This chapter considers the recognition of First Nations[1] peoples' rights and obligations in sustainable water management and use in a just transition towards decarbonisation, with a focus on Australia. The authors begin by noting we are not Indigenous. However, our aim here is to highlight how non-Indigenous water law and policy affects First Nations people and by doing so, help to facilitate a space for First Nations voices to be heard so they can contribute to the transition towards decarbonisation and ensure the transition for them is a just one.

Transitioning to decarbonisation is essential for tackling climate change. Under the United Nations Climate Change Convention's Conference of the Parties (COP) theme of nature, this requires the sustainable management, adaptation and use of water resources given the effects of climate change on water resources and the role that water plays in—among other things—land management, agriculture and emerging decarbonising technologies.[2] Accordingly, we need to harness the collective knowledge of water management and use to ensure the transition is not only sustainable, but also just. Knowledge of water resource management and use that has been largely overlooked to date, particularly in legislative frameworks, is that of First Nations people.

For First Nations people, water has meaning beyond simply drinking, domestic, agricultural and environmental purposes; it also has a vital cultural and spiritual component.[3] Further, water cannot be separated

[1] The terms 'First Nations' or 'Indigenous' will be used throughout this chapter unless the context requires otherwise. There is no consistent terminology in Australian law and policy, which refers variously to Indigenous people, Aboriginal people, Aboriginal and Torres Strait Islander people, Traditional Owners, First Nations and First Peoples. Indigenous peoples is the term used in the international domain, e.g., United Nations Declaration on the Rights of Indigenous Peoples.

[2] For example, the processing of rare earth minerals: 'Processing the raw ore into a form useful to make magnets and other tech is a lengthy effort that takes large amounts of water'. Carolyn Gramling, 'Rare Earth Mining May be Key to our Renewable Future. But at What Cost?' *Science News* (Online, 11 January 2023). https://www.sciencenews.org/article/rare-earth-mining-renewable-energy-future.

[3] See, e.g., Marcia Langton, 'Earth, Wind, Fire and Water: The Social and Spiritual Construction of Water in Aboriginal Societies', in Bruno David, Bryan Barker and Ian J. McNiven (eds), *The Social Archaeology of Australian Indigenous Societies* (Aboriginal Studies Press, 2006) 139.

from land; all is considered 'Country'.[4] First Nations people have a wealth of knowledge about land and water management and use, based on thousands of years of continuing use and adaptation.[5] In relation to Australia, it has been noted that:

> Indigenous Australians have managed their lands and waters sustainably for thousands of generations. Through their spiritual, cultural and customary connections to the landscape, they have acquired a deep knowledge and understanding of Australia's water systems.[6]

Indigenous people in Australia are also disproportionally affected by climate change. According to Nikolakis et al., '[Indigenous peoples] are more vulnerable to the impacts from climate change as a result of their location on marginal lands and socio-economic disadvantage, which constrains their ability to adapt to change'.[7] Climate change not only impacts on their access to safe drinking water, it also negatively affects their cultural and spiritual connection to Country. As noted in Australia's *2021 State of the Environment Report*, 'changes [to aquatic ecosystems caused by climate change] will affect Indigenous people the most, given their spiritual connection with the environment'.[8]

[4] There are multiple definitions of 'Country'; here is but one: 'Country is a worldview that encompasses our relationship to the physical, ancestral and spiritual dimensions, and involves the kind of intimacy evident in the oft quoted expression "The Country is our mother. We belong to the country; it does not belong to us"'. Bill Gammage and Bruce Pascoe, *Country: Future Fire, Future Farming* (Thames & Hudson, 2021) 5.

[5] See, e.g., *Our Common Future: Report of the World Commission on Environment and Development*, UN GA, 42nd sess, UN Doc A/42/427 (4 August 1987) annex 119 para 74; *AR5 Climate Change 2014: Impacts, Adaptation, and Vulnerability*, Working Group II Contribution to the Fifth Assessment Report of the IPCC ('Climate Change 2014') 758.

[6] National Water Commission, *Position Statement: Indigenous Access to Water Resources* (June 2012) 1.

[7] William Nikolakis, Quentin Grafton and Aimee Nygaard, 'Indigenous Communities and Climate Change: A Recognition, Empowerment and Devolution (RED) Framework in the Murray–Darling Basin, Australia' (2016) 7(1) *Journal of Water and Climate Change* 169, 171; see also Donna Green, Sue Jackson and Joe Morrison (eds), *Risks from Climate Change to Indigenous Communities in the Tropical North of Australia* (Commonwealth Department of Climate Change, 2009) 18.

[8] Australia, State of the Environment 2021: Inland Water, 12.

Further, the building of dams for water security or hydro power affects water sources and impacts Indigenous Country. This is not a new phenomenon,[9] but is exacerbated by climate change. Rising sea levels affect Pacific (Indigenous) nations and Torres Strait Islander peoples.[10]

On a global scale, various Working Group II (WGII) reports to the Intergovernmental Panel on Climate Change (IPCC) have repeatedly noted the impact of climate change on Indigenous peoples.[11] For example, the most recent such report states that, '[c]limate change is impacting Indigenous Peoples' ways of life ..., cultural and linguistic diversity ..., food security ... and health and well-being'.[12]

Western/non-Indigenous laws should therefore recognise the value of Indigenous knowledge, to enable First Nations people to maintain their spiritual and cultural connection to, and responsibility for, Country. Such knowledge is also vital for the effective adaptation of water resource management and use. Both aspects were noted in the WGII 2022 report:

> Indigenous knowledge and local knowledge can contribute to reducing the vulnerability of communities to climate change .... Supporting Indigenous self-determination, recognising Indigenous Peoples' rights and supporting Indigenous knowledge-based adaptation are critical to reducing climate change risks and effective adaptation.[13]

The implementation plan from COP27 held in Egypt in 2022 reinforces this in recognising 'the important role of Indigenous people ...

[9] The Tasmanian Dams case involved protection of Aboriginal sites that were in danger from the construction of the dam: *Commonwealth v Tasmania (1983)* 158 CLR 1.

[10] Donna Green et al., 'An Assessment of Climate Change Impacts and Adaptation for the Torres Strait Islands, Australia' (2010) 102 *Climate Change* 405.

[11] See, e.g., *Climate Change 2007: Impacts, Adaptation and Vulnerability*, Working Group II Contribution to the Fourth Assessment Report of the Intergovernmental Panel on Climate Change, 522–23; *AR5 Climate Change 2014: Impacts, Adaptation, and Vulnerability*, Working Group II Contribution to the Fifth Assessment Report of the IPCC, 77 ('Climate Change 2014').

[12] *Climate Change 2022: Impacts, Adaptation and Vulnerability*, Working Group II Contribution to the Sixth Assessment Report of the Intergovernmental Panel on Climate Change, 53 ('Climate Change 2022').

[13] Ibid., 53.

in addressing and responding to climate change'.[14] The implementation plan also 'emphasizes the importance of protecting, conserving and restoring water and water-related ecosystems, including river basins, aquifers and lakes',[15] as part of the response to climate change.

In keeping with the just transitions theme of this book, this chapter revolves around the three commonly recognised principles of justice that are considered to form the basis of a just transition: distributive justice, procedural justice and recognition justice. As explained by Krawchenko and Gordon:

> Within a just transitions approach, justice can be conceived in distributional, procedural, and recognitional terms: distributional justice is concerned with how different groups benefit or experience impacts from the changes required; recognitional justice identifies interest groups and rights holders who may be implicated; and procedural justice is concerned with elements of governance—who is included and how.[16]

In a First Nations water justice context, these three types of justice can be described as follows. Distributive justice is concerned with First Nations relative level of access to water resources for self-determined purposes; procedural justice is concerned with First Nations degree of engagement and participation in governance structures and processes; and recognition justice is concerned with First Nations ability to have cultural values and worldviews associated with water resources reflected in distributional rules and governance structures and processes.

The focus of this chapter is Australian water law and policy, for several reasons. First, Australia is regarded as the driest inhabited continent on Earth, with extreme variability in rainfall and runoff, both spatially

[14] Decision -/CP.27: *Sharm el-Sheikh Implementation Plan* (advance unedited version) Pt XVI art 51.

[15] Ibid Pt V art 21.

[16] Tamara Antonia Krawchenko and Megan Gordon, 'How Do We Manage a Just Transition? A Comparative Review of National and Regional Just Transition Initiatives' (2020) 13(11) *Sustainability* 6070: 1–16, 1.

and temporally[17] and only getting drier because of climate change.[18] Therefore, how Australia manages and uses its water resources is of vital importance. Second, it is commonly understood that First Nations Australians have the longest continuing living culture on Earth[19] and thus have water knowledge extending back many thousands of years.[20] However, it was not until 1992 when the High Court of Australia handed down its landmark decision in *Mabo v Queensland [No 2]*[21] that the concept of terra nullius (land belonging to no one) was rejected and First Nations rights and interests in land and waters could be formally recognised. Further, the WGII has noted that 'Indigenous peoples in both Australia and New Zealand have higher than average exposure to climate change due to a heavy reliance on climate-sensitive primary industries and strong social connections to the natural environment'.[22] Along with a complex water management regime, these make Australia a useful case study, providing potential lessons for other First Nations peoples and nation states.

This chapter begins by outlining Australia's water management regime. It then puts a spotlight on Commonwealth water law and policy to evaluate the scope these provide for First Nations participation in water management; the recognition they accord to First Nations rights and interests; and the water use rights they give to First Nations Australians, all of which are necessary for a just transition to decarbonisation. It

[17] David Ingle Smith, *Water in Australia: Resources and Management* (Oxford University Press, 1998) 4–16; see also Will Steffen, *Thirsty Country: Climate Change and Drought in Australia* (Report, Climate Council of Australia, 25 March 2015) 2. https://apo.org.au/node/53961.

[18] See, e.g., Robert David Pilz, 'Lessons in Water Policy Innovation from the World's Driest Inhabited Continent: Using Water Allocation Plans and Water Markets to Manage Water Scarcity' (2010) 14(1) *University of Denver Water Law Review* 97, 100–101; see also Steffen (Footnote 17).

[19] See, e.g., Melissa Perry, 'The Law, Equality and Inclusiveness in a Culturally and Linguistically Diverse Society' (2019) 40(1) *Adelaide Law Review* 273, 274.

[20] Evidence of human occupation in Australia extends back 65,000 years: Chris Clarkson et al., 'Human Occupation of Northern Australia by 65,000 Years Ago' (2017) 547 *Nature* 306.

[21] (1992) 175 CLR 1 ('*Mabo [No 2]*').

[22] Climate Change 2014 (Footnote 11) 77.

then evaluates state and territory water law and policy through the same lens and concludes with some observations on the capacity of Australia's water management regime to deliver justice to First Nations people in the transition to decarbonisation.

4.2 Australia's Water Management Regime and the Role of Indigenous People

Australia has a federal system of government established by the *Australian Constitution* ('*Constitution*'). Under the *Constitution*, the Commonwealth Parliament is given the power to legislate in relation to a list of enumerated topics, power that it holds concurrently with the states.[23] That list makes no reference to water or the environment. The only specific roles the Commonwealth Parliament has in relation to water are found in s 98 of the *Constitution*, which relates to navigation and shipping, and s 100 whereby the Commonwealth cannot 'by law or regulation of trade or commerce, abridge the right of a State or its residents therein to the reasonable use of waters of rivers for conservation or irrigation'.[24]

Consequently, water management and use are largely governed by state legislation. Nonetheless, the Commonwealth Parliament has been able to use its enumerated constitutional powers to pass some water and environmental legislation.[25] Of particular relevance here are the *Water Act 2007* (Cth) ('*Water Act*') and the *Environment Protection and Biodiversity Conservation Act 1999* (Cth) ('*EPBC Act*'), both of which are discussed below. Another piece of Commonwealth legislation relevant to, or impacting on, First Nations water management and use discussed here is the *Native Title Act 1993* (Cth) ('*NTA*'), which sets out the process by which native title rights and interests in land and waters

[23] *Australian Constitution*, s 51. The Commonwealth Parliament has some exclusive legislative powers in s 52.

[24] Kildea and Williams note that at Federation, the term conservation referred to the 'impounding of water'. They go on to say, 'it seems arguable that it should be given a broader construction consistent with its contemporary meaning'. Paul Kildea and George Williams, 'The Constitution and the Management of Water in Australia's Rivers' (2010) 32 *Sydney Law Review* 595, 610.

[25] In particular, s 51(xxix) the external affairs power. The *Water Act 2007* (Cth) ('*Water Act*') also involved the referral power in s 51 (xxxvii), which allows the states to refer power to the Commonwealth.

can be recognised.[26] Finally, the Commonwealth has also developed, by way of agreement with all Australian states and territories, a national water policy—namely the *National Water Initiative* ('*NWI*').[27] This is discussed first.

4.3 Commonwealth Water Law and Policy

4.3.1 The 2004 National Water Initiative

The 2004 *NWI* is the leading Australian national policy document governing water management and use. It was described at the time of its development as 'represent[ing] the most significant innovation in water policy in more than a century'.[28] Notably, it is the first national policy document since *Mabo [No 2]*[29] and the *NTA* to recognise Indigenous water rights.[30] Those rights, however, are quite limited in scope, relating mostly to native title rights and being largely aspirational in their terms. The main provisions in that regard are set out below (emphasis added):

> 52. The Parties will provide for indigenous access to water resources, in accordance with relevant Commonwealth, State and Territory legislation, through planning processes that ensure:
>
> i) inclusion of indigenous representation in water planning <u>wherever possible</u>; and
> ii) water plans will incorporate indigenous social, spiritual and customary objectives and strategies for achieving these objectives <u>wherever they can be</u> developed.

[26] Cultural heritage protection legislation—that is, the *Aboriginal and Torres Strait Islander Heritage Protection Act 1984* (Cth) and corresponding state and territory legislation—is also relevant because water sites are generally regarded as culturally significant. In Victoria, for example, waterways are designated areas of cultural heritage sensitivity: *Aboriginal Heritage Regulations 2018* (Vic), reg 26. However, an extensive discussion of cultural heritage legislation is beyond the scope of this chapter.

[27] *Intergovernmental Agreement on a National Water Initiative* (24 June 2004).

[28] Daniel Connell, Stephen Dovers and R Quentin Grafton, 'A Critical Analysis of the National Water Initiative' (2005) 10(1) *Australasian Journal of Natural Resources Law and Policy* 81, 85.

[29] (1992) 175 CLR 1.

[30] Katie O'Bryan, *Indigenous Rights and Water Resource Management: Not Just Another Stakeholder* (Routledge, 2019) 47, 63 ('Not Just Another Stakeholder').

> 53. Water planning processes will take account of the possible existence of native title rights to water in the catchment or aquifer area. The Parties note that plans may need to allocate water to native title holders following the recognition of native title rights in water under the Commonwealth Native Title Act 1993.
>
> 54. Water allocated to native title holders for traditional cultural purposes will be accounted for.

Various other provisions also require recognition or consideration of Indigenous interests.[31] Finally, Indigenous and cultural values fall under the definition of 'environmental and other public benefit outcomes'.[32]

The *NWI* as it relates to Indigenous interests has been the subject of much criticism.[33] In terms of contributing towards a just transition for First Nations Australians, the relevant elements of the *NWI* amount to very weak forms of justice, given their aspirational nature and limited scope.

Importantly, the *NWI* was the subject of a review by the Commonwealth Productivity Commission, the final report of which was released in 2021.[34] Although the Productivity Commission concluded that all jurisdictions had made good progress in implementing the reform agenda in the *NWI*, it noted that progress in improving the scale and quality of engagement with Aboriginal and Torres Strait Islander people had been slow.[35] It therefore recommended that a new element of the *NWI* be

[31] Other provisions referring to Indigenous use and significance are cls 25(ix), sch E cl(vi), sch G cl 3(v).

[32] 2004 *National Water Initiative* sch B(i) (definition of 'environmental and other public benefit outcomes').

[33] See, e.g., Virginia Falk, 'Beyond Native Title: The "Rights" Response to Social Justice and the Conceptual Problem of Resource Development' (Conference Paper, National Native Title Conference, 1–3 June 2010, Canberra); Sue Jackson, *Indigenous Interests and the National Water Initiative* (Report, 2007) 64; Martuwarra RiverOfLife, Katherine S Taylor and Anne Poelina, 'Living Waters, Law First: Nyikina and Mangala Water Governance in the Kimberly, Western Australia' (2021) 25(1) *Australasian Journal of Water Resources* 40.

[34] Productivity Commission, *National Water Reform 2020: Inquiry Report* (Report No 96, 28 May 2021).

[35] Ibid 1–2.

developed covering Aboriginal and Torres Strait Islander people's interests in water.[36] To that end, a Committee on Aboriginal and Torres Strait Islander Water Interests comprised of Aboriginal and Torres Strait Islander people has been established, with one of its tasks being to progress this new element.[37] The Productivity Commission also emphasised the need for climate change to be taken into account in water planning.[38] Both are inherent in the following statement:

> [T]he renewed NWI should also explicitly mention the need to adapt to a changing climate, and recognise Aboriginal and Torres Strait Islander people's reverence and cultural responsibility for rivers and groundwater systems, and their desire to participate in all significant water-related processes and decisions.[39]

This is a clear acknowledgement of the need for procedural justice for First Nations Australians.

At the time of writing, the Commonwealth Government had not yet formally responded to the Productivity Commission's report as it was waiting on the outcome of a consideration of the report by the National Water Reform Committee.[40]

The Commonwealth Government has also developed a First Nations water policy in which it states that it is 'working with First Nations Australians to increase ownership of water and to take part in water policy decisions', and commits to several initiatives in that regard.[41] This presents an opportunity for both distributive and procedural justice outcomes for First Nations Australians.

[36] Ibid 121.

[37] *Committee on Aboriginal and Torres Strait Islander Water Interests* (Web Page, 23 August 2023). https://www.dcceew.gov.au/water/policy/first-nations/cawi.

[38] Productivity Commission (Footnote 34) 6.

[39] Ibid., 6.

[40] *National Water Initiative* (Web Page, 9 December 2022). https://www.dcceew.gov.au/water/policy/policy/nwi; for more on the composition and role of the National Water Reform Committee, see *National Water Policy* (Web Page, 12 July 2023). https://www.dcceew.gov.au/water/policy/policy.

[41] *First Nations Water Policy* (Web Page, 7 July 2023). https://www.dcceew.gov.au/water/policy/first-nations.

The Productivity Commission's review and report referred to above were statutory requirements under the *Water Act*,[42] to which we now turn.

4.3.2 *The* Water Act

The *Water Act* is the main Commonwealth statute relating to water. It relates largely, although not exclusively, to the management of the Murray–Darling Basin (MDB). The MDB extends across four states and one territory—Queensland, New South Wales (NSW), Victoria, South Australia and the Australian Capital Territory (ACT)—and is 'Australia's most important water catchment'.[43]

The *Water Act* recognises the important role of Indigenous people in water management and use by providing a mandated Indigenous position on the Murray–Darling Basin Authority (MDBA), the entity that prepares, implements and monitors the *Murray–Darling Basin Plan* ('*Basin Plan*').[44] The *Basin Plan* is a legislative instrument the purpose of which is 'to provide for the integrated management of the Basin water resources'.[45] To be eligible for appointment, an Indigenous person must have a high level of expertise in 'Indigenous matters relevant to Basin water resources'.[46]

The *Water Act* also establishes the Basin Community Committee (BCC), composed of up to 17 members (including the chair), which must include at least two Indigenous representatives 'with expertise in Indigenous matters relevant to the Basin's water resources'.[47] At the time of writing there were three members on the BCC who identify as Indigenous.[48]

[42] *Water Act* s 88.

[43] *Basin Location* (Web Page, 15 June 2023). https://www.mdba.gov.au/basin/basin-location.

[44] *Water Act* s 177(b).

[45] Ibid., s 20.

[46] Ibid., s 178(2A).

[47] Ibid., s 202(5)(c).

[48] Phil Duncan, Feli McHughes and Alice Williams: *Basin Community Committee* (Web Page, 28 July 2023). https://www.mdba.gov.au/about-us/governance-and-committees/basin-community-committee.

These provisions evidence forms of procedural justice, being engagement and participation in governance structures and processes. The stronger of the two participatory mechanisms is the dedicated Indigenous position on the MDBA, because of its important governance functions, including in relation to the *Basin Plan*. In contrast, the BCC is merely an advisory body.

The *Basin Plan* prepared by the MDBA must include information about 'the uses to which the Basin water resources are put (including by Indigenous people)'.[49] Further, a water resources plan,[50] required by the *Basin Plan*, must be prepared 'having regard to social, spiritual and cultural matters relevant to Indigenous people in relation to the water resources of the water resource plan area in the preparation of the water resource plan'.[51]

These are forms of recognition justice because they identify First Nations as relevant water users that may be impacted and require that First Nations cultural values and world views associated with water are considered. However, merely requiring that First Nations values and world views are considered does not mean that this will result in a substantive outcome.

4.3.3 The EPBC Act

The *EPBC Act* is the Commonwealth's premier piece of environmental legislation. The subject of two statutory reviews,[52] the *EPBC Act* is viewed as ineffective, inefficient and in need of fundamental reform

[49] Ibid., s 22(1).

[50] These are legally enforceable plans made by the states and territories.

[51] *Water Act* s 22(3)(ca). For more on Indigenous interests in the 2012 *Murray–Darling Basin Plan* itself, see O'Bryan, *Not Just Another Stakeholder* (Footnote 30) 90–94.

[52] Graeme Samuel, *Independent Review of the EPBC Act—Final Report* (October 2020) ('Samuel Review'). The first statutory review of the *EPBC Act* was undertaken in 2008–9, the final report of which was released in October 2009: Allan Hawke, *The Australian Environment Act: Report of the Independent Review of the Environment Protection and Biodiversity Conservation Act 1999* (Department of the Environment, Water, Heritage and the Arts, October 2009).

(about which more is said below).[53] Nonetheless, it also has some relevance to the management and use of water resources.[54] However, the *EPBC Act* is only triggered in relation to matters of national environmental significance. Insofar as these matters may relate to water, they include World Heritage properties,[55] National Heritage places,[56] Wetlands of international importance[57] and the protection of water resources from coal seam gas development and large coal mining development.[58]

The *EPBC Act* clearly contemplates a role for Indigenous people, as evidenced in the objects of the Act[59] and how those objects are achieved, namely by promoting 'a partnership approach to environmental protection and biodiversity conservation through ... recognising and promoting indigenous peoples' role in, and knowledge of, the conservation and ecologically sustainable use of biodiversity'.[60]

To facilitate this, the *EPBC Act* establishes an Indigenous Advisory Committee (IAC), the purpose of which 'is to advise the Minister on the operation of the Act, taking into account the significance of indigenous peoples' knowledge of the management of land and the conservation and sustainable use of biodiversity'.[61] However, as the name suggests, it is advisory only, and its advice relates to the operation of the *EPBC Act* in its entirety (i.e. not limited to water) and at a strategic level. It also officially meets only twice a year (although it does have regular teleconferences and out-of-session work).[62] Thus, there are significant practical

[53] Samuel Review, 24–25.

[54] For further detail on the *Environment Protection and Biodiversity Conservation Act 1999* (Cth) ('*EPBC Act*') and First Nations water management, see O'Bryan, *Not Just Another Stakeholder* (Footnote 30) 94–100.

[55] *EPBC Act* ss 12–15A.

[56] Ibid., ss 15B–15C.

[57] Ibid., ss 16–17B.

[58] Ibid., ss 24D–24E.

[59] Ibid., s 3(1)(d)(f) and (g).

[60] Ibid., s 3(2)(g)(iii).

[61] *EPBC Act* s 505B(1).

[62] Ibid.

limitations to its effectiveness.[63] Accordingly, although there is important recognition of Indigenous peoples in the objects of the *EPBC Act* (recognition justice), in terms of procedural justice (i.e. involvement in governance structures), it is quite limited.

Recognition justice also manifests in various other forms in the *EPBC Act*, in facilitation of the Act's objectives; for example, by making it an offence to take an action in relation to a National Heritage place that will result in a significant impact on Indigenous heritage values.[64] Examples of National Heritage places relevant to water with significant Indigenous heritage values include the Brewarrina Aboriginal Fish Traps (Baiame's Ngunnhu) in the Darling River in NSW[65] and the Budj Bim National Heritage Landscape at Lake Condah in Victoria.[66] Notably, Budj Bim is also a World Heritage property.[67] There are also various requirements to consider the role of Indigenous peoples; for example, when the Minister is entering into a bilateral agreement with a state.[68]

Like the *NWI*, the *EPBC Act* has recently been the subject of a statutory review, the report of which was released in October 2020 (the 'Samuel Review').[69] The Samuel Review recognised that 'Indigenous knowledge and views are not fully valued in decision-making'.[70] A number of recommendations were made to address this concern. In particular, Recommendation 5 states that, '[t]o harness the value and recognise the importance of Indigenous knowledge, the EPBC Act should require decision-makers to respectfully consider Indigenous views and knowledge'.[71] One of the reforms recommended in that regard is

[63] For more on the Indigenous Advisory Committee, see O'Bryan, *Not Just Another Stakeholder* (Footnote 30) 95–96.

[64] *EPBC Act* s 15C(7) and (8).

[65] *National Heritage Places—Brewarrina Aboriginal Fish Traps* (Baiame's Ngunnhu) (Web Page, 30 October 2021). https://www.dcceew.gov.au/parks-heritage/heritage/places/national/brewarrina.

[66] *National Heritage Places—Budj Bim National Heritage Landscape* (Web Page, 3 October 2021). https://www.dcceew.gov.au/parks-heritage/heritage/places/national/budj-bim.

[67] *Budj Bim Cultural Landscape* (Web Page). https://whc.unesco.org/en/list/1577.

[68] *EPBC Act* s 49A(c).

[69] Samuel Review (Footnote 52).

[70] Ibid., 6.

[71] Ibid., 27.

to replace the IAC with an Indigenous Engagement and Participation Committee, the mandate of which would be 'to refine, implement and monitor the National Environmental Standard for Indigenous engagement and participation in decision-making'.[72] These recommendations, if implemented, would enhance procedural justice for First Nations.

The government responded to the Samuel Review with its *Nature Positive Plan*.[73] One of the commitments in this plan is to '[work] in partnership with First Nations to achieve better outcomes for people and communities and for Australia's heritage and environment'.[74] 'Working in partnership' suggests a procedural justice approach, which may result in distributive and recognition justice outcomes.

4.3.4 *The Native Title Act*

As noted earlier, First Nations input into water management and use may occur pursuant to legislation relating specifically to First Nations people, the focus here being the *NTA*.

The *NTA* provides for recognition by the courts of native title rights and interests over land and waters held pursuant to traditional laws and customs.[75] As of 1 July 2023, native title had been recognised over approximately 43% of Australia.[76] The *NTA* also establishes a 'future act' regime that regulates activities that affect (or may affect) native title. This includes consultation procedures ranging from the right to negotiate[77] to a right to be notified with an associated opportunity to comment.[78]

[72] Samuel Review 27.

[73] Department of Climate Change, Energy, the Environment and Water, *Nature Positive Plan: Better for the Environment, Better for Business* (December 2022).

[74] Ibid., 13.

[75] See *Native Title Act 1993* (Cth) ('*NTA*') s 223 for the definition of native title.

[76] This figure has been calculated using data from the National Native Title Tribunal National Map: *Native Title Determinations: 1 July 2023, Maps* (Web Page), http://www.nntt.gov.au/assistance/Geospatial/Pages/Maps.aspx and Geosciences Australia, *Areas of Australia—States and Territories* (Web Page, 26 July 2023). https://www.ga.gov.au/scientific-topics/national-location-information/dimensions/area-of-australia-states-and-territories.

[77] *NTA* ss 25–44.

[78] See, e.g., *NTA* ss 23HA, 24GB, 24GD, 24GE, 24HA, 24ID, 24JAA, 24JB.

The management of water falls into the latter category.[79] A failure to notify, however, has been interpreted by the courts as not invalidating an activity relating to the management of water resources.[80] This renders the ability of native title holders (and claimants) to have any meaningful input almost nugatory, as there is little incentive for the decision-maker to follow those notification procedures.[81] Accordingly, in terms of procedural justice (participation in decision-making or governance structures), the future act regime is very weak.

There may, however, be some scope for input into water management by native title holders in areas where they have been recognised as having exclusive possession. This is because exclusive possession can amount to control of access to land by those native title holders, which in turn may enable some control over water on that land.[82]

However, recognition of exclusive possession native title to date has occurred mostly in Western Australia, where approximately 93% of all exclusive possession native title is found.[83] There is limited potential for its recognition elsewhere, particularly in the more settled regions of Australia where colonial dispossession has had its greatest impact. Accordingly, the *NTA* does not provide much assistance to First Nations people to have meaningful input into the management of water.

In relation to the recognition of native title rights to take and use water (in contrast to management of water), in the past this has generally been limited to the right to take and use for essentially cultural, spiritual and domestic non-commercial communal needs. However, following the 2013 High Court decision in *Akiba v Commonwealth* ('*Akiba*'),[84] there is the potential for expansion of native title rights to take and use water. In *Akiba*, the High Court held that there was a right to take resources for any purpose, which could include for commercial purposes. At the time of writing there has been no judicial determination on this point

[79] *NTA* s 24HA.

[80] *Lardil Peoples v Queensland* (2001) 108 FCR 453, 471–2, 473 (French J), 486–7 (Dowsett J).

[81] For further detail on the *NTA* and the management of water, see O'Bryan, *Not Just Another Stakeholder* (Footnote 30) 69–88.

[82] See *Gumana v Northern Territory [No 2]* [2005] FCA 1425 [43].

[83] This figure has been calculated from data on the National Native Title Tribunal map of Native Title Determinations (Footnote 76).

[84] (2013) 250 CLR 209.

in relation to water (as a resource). In light of *Akiba*, some exclusive possession consent determinations no longer contain the qualification that water access and use be only for cultural, spiritual and domestic non-commercial communal needs.[85]

In the context of a just transition, and in particular distributive justice, for justice to have any meaning requires that First Nations native title rights to take and use water be recognised and preserved for all purposes, not merely for spiritual, cultural and domestic, non-commercial needs.

There is also the possibility of a management role in relation to water sites of cultural heritage significance,[86] given that most (if not all) positive determinations of native title include a right to protect and maintain cultural heritage sites.[87] Cultural heritage protection does not rely on a determination of native title, so in that respect, it has advantages for First Nations people who are not native title holders, or who may not meet the high threshold for recognition of native title under the *NTA*. However, in practice the protection of First Nations cultural heritage relies on effective cultural heritage legislation.

In that respect, the ineffectiveness of Australia's cultural heritage protection legislation was exposed when, on 24 May 2020, Rio Tinto destroyed Juukan Gorge, a significant site dating back 46,000+ years. The destruction of Juukan Gorge precipitated a parliamentary inquiry, the final report of which was released in October 2021.[88] The recommendations arising out of that report included a new federal cultural heritage legislative framework to be co-designed with Aboriginal and Torres Strait Islander peoples which should set out minimum standards for state and territory heritage protection laws.[89]

[85] See, e.g., *Minning v Western Australia* [2020] FCA 1051 O 7(b).

[86] See, e.g., Michael O'Donnell, *NAILSMA–TRaCK Project 6.2: Indigenous Rights in Water in Northern Australia* (March 2011) 127–32.

[87] See, e.g., *Mullett v Victoria* [2010] FCA 1144 O 5(d), which describes this as the 'right to protect and maintain places and areas on the land and waters which are of importance according to Gunai/Kurnai traditional laws and customs'.

[88] Joint Standing Committee on Northern Australia, Parliament of Australia, *A Way Forward: Final Report into the Destruction of Indigenous Heritage Sites at Juukan Gorge* (Final Report, October 2021).

[89] Ibid., Recommendation 3.

4.4 State and Territory Water Law and Policy

4.4.1 *First Nations and State and Territory Water Laws*

As noted earlier, water management and use are largely regulated by the states. The ACT and the Northern Territory (NT) also have their own laws governing water management and use. However, although many state and territory water laws and policies now refer to First Nations people, there is a wide divergence in the level of legislative recognition and protection of First Nations water knowledge, rights and interests.

For example, in Victoria, First Nations people have a right to take and use water if they have an agreement under the *Traditional Owner Settlement Act 2010* (Vic) ('*TOS Act*'), Victoria's equivalent of the *NTA*.[90] There are also requirements for Water Authorities and Catchment Management Authorities to give notice to or consult with specified Aboriginal parties (i.e. native title holders, traditional owner group entities and registered Aboriginal parties) in relation to the preparation of various plans and strategies,[91] and to include an Aboriginal person on certain consultative committees.[92] Further, the Victorian Environmental Water Holder, Water Corporations and Sustainable Water Strategies must 'consider opportunities to provide for ... Aboriginal cultural values and uses of waterways'.[93]

This can be contrasted with the ACT's *Water Resources Act 2007*, which makes no reference to First Nations people. Tasmania's *Water Management Act 1999* makes only one reference to Aboriginal people,[94] as does Western Australia's primary water law, the *Rights in Water and Irrigation Act 1914* (WA).[95]

Legislation in each of NSW, Victoria, Queensland and South Australia has objects or purposes that promote recognition of First Nations cultural

[90] *Water Act 1989* (Vic) s 8A.

[91] Ibid., ss 22E, 22M, 22R, 31, 190. *Catchment and Land Protection Act 1994* (Vic) sch 2 cl 2(3)(ba).

[92] *Water Act 1989* (Vic) ss 22D, 22Q, 29.

[93] Ibid., s 22C(3)(a)(i); s 33DD(2)(a); s 92(2A)(a); s 189(2)(a).

[94] Section 193(i).

[95] Section 26GL(1)(b)(iii). Western Australia's water regime currently consists of six laws, of which the *Rights in Water and Irrigation Act 1914* (WA) ('*RIWI Act*') is the most relevant.

and spiritual values and uses of water.[96] For example, one of the purposes of Victoria's *Water Act 1989* is 'to ensure that Victoria's water resources and waterways are managed in a way that considers … Aboriginal cultural values and uses of waterways'.[97]

These are manifestations of recognition justice. However, they are largely symbolic, and do not guarantee any meaningful outcomes for First Nations people, as it is not possible to 'enforce' an objects clause. Recognition needs to be accompanied by the ability to influence decisions and/or have a role in water governance. It is only by involving First Nations people in governance and decision-making that their knowledge can properly inform (and their rights and interests be adequately protected in) the adaptations needed to transition to decarbonisation.

In that respect, some state laws provide for positions for First Nations people (or people with qualifications or experience relating to First Nations people) on management committees,[98] landscape boards[99] or advisory committees.[100] These entities have functions ranging from the preparation of plans and strategies to the provision of advice.

A just transition also requires First Nations people to be able to access and use water, for any purpose. Again, there is wide divergence among the states and territories. Following is a brief outline of those provisions.

NSW has a 'specific purpose access licence' that can be used for Aboriginal cultural or commercial purposes.[101] There is a fee exemption for Aboriginal commercial, Aboriginal community development and Aboriginal cultural purposes.[102] In Victoria, as noted above, First Nations people can access and use water under s 8A of the *Water Act 1989* if they have an agreement under the *TOS Act*. However, only a limited number of First

[96] *Water Management Act 2000* (NSW) s 3(c)(iv); *Water Act 1989* (Vic) s 1(ea)(i); *Water Act 2000* (Qld) s 2(2)(d); *Landscapes South Australia Act 2019* (SA) s 7(1)(b); (3)(a) and (e).

[97] *Water Act 1989* (Vic) s 1(ea)(i).

[98] See, e.g., *Water Management Act 2000* (NSW) s 13; *RIWI Act* s 26GL(1)(b)(iii).

[99] See, e.g., *Landscape South Australia Act 2019* (SA) s 16(1)(i).

[100] See, e.g., *Water Act 1989* (Vic) s 22D(2)(b), s 22Q(2)(a)(iv); see also *Water Act 1992* (NT) s 24(3)(b)(iv).

[101] Aboriginal licences are sub-categories of various access licences: *Water Management (General) Regulations 2018* (NSW) sch 3.

[102] *Fees and Application Forms* (Web Page). https://water.dpie.nsw.gov.au/licensing-and-trade/fees-and-application-forms.

Nations groups have a *TOS Act* agreement,[103] and the use to which the water can be put is restricted to traditional purposes[104] as per the *TOS Act*.[105]

Queensland and the NT have a system of Aboriginal water reserves, described as '[w]ater reserves to meet the social and economic aspirations of Indigenous communities'.[106] In Queensland, Aboriginal and Torres Strait Islander people can also 'take or interfere with water for traditional activities or cultural purposes'.[107] This does not, however, include commercial purposes.[108] The Western Australian, South Australian and Tasmanian water laws are silent on water access and use allocations for First Nations people. Western Australia, however, is currently reviewing its water management regime. The bill (still to be released) will seek to 'ensure Aboriginal land, heritage, economic and cultural development concerns are effectively represented in water policy development and water allocation processes' and 'deliver benefits to Aboriginal people in relation to their spiritual, social, customary and economic use of water'.[109]

This brief survey of state and territory water legislation indicates that it is currently inadequate to ensure a just transition for First Nations people, as the quality of justice is dependent on the extent to which First Nations knowledge, rights, interests and obligations to Country have been accommodated in each state's or territory's water law.

[103] Currently the Gunaikurnai, Dja Dja Wurrung and Taungurung peoples.

[104] 'Traditional purposes' is defined in s 79 of the *Traditional Owner Settlement Act 2010* (Vic) ('*TOS Act*') and, in summary, is for personal, domestic and non-commercial communal needs.

[105] *TOS Act* s 82, in conjunction with s 84.

[106] Lee Godden, Sue Jackson and Katie O'Bryan, 'Indigenous Water Rights and Water Law Reforms in Australia' (2020) 37 *Environmental and Planning Law Journal* 655, 672.

[107] *Water Act 2000* (Qld) s 95(1).

[108] Ibid., s 95(2).

[109] *Water Reform Bill* (Web Page, 24 October 2022). https://www.wa.gov.au/service/building-utilities-and-essential-services/water-supply/water-reform-bill.

4.4.2 First Nations and State and Territory Water Policy

Despite the patchwork of recognition in water law across Australia, as demonstrated above, several states and territories refer to First Nations people in their state- or territory-wide water policy.[110] Some states and territories also have, or are developing, a specific First Nations water policy.

In that respect, Victoria released its First Nations water policy on 29 September 2022.[111] Called *Water is Life: Traditional Owner Access to Water Roadmap*,[112] the policy has two sections. Section A contains the Victorian Government policy position. Section B consists of water policy statements made by Traditional Owner Nations.[113] The inclusion of Section B is important because it contains the vision of Traditional Owner Nations on water justice.

NSW is in the process of developing a First Nations water policy,[114] making good on its commitment in the *NSW Water Strategy: Towards 2050*.[115] So, too, is Queensland, as part of the government's *Path to Treaty*.[116] In the ACT, following a Water Governance Review in 2021–22, the ACT Office of Water committed to delivering 'a refreshed water

110 *Water for Victoria: Water Plan* (Department of Environment, Land, Water and Planning, 2016), ch 6. https://www.water.vic.gov.au/about-us/water-for-victoria; *NSW Water Strategy*, Priority 2 (Web Page). https://water.dpie.nsw.gov.au/plans-and-programs/nsw-water-strategy/toward-2050; *ACT Water Strategy 2014-44: Striking the Balance* (ACT Government: Environment and Planning, August 2014), part 3.3 and strategy 7 (action 18).

111 Harriet Shing, 'Water is Life—Part of Victoria's Water Future' (Media Release, 29 September 2022).

112 (Department of Environment, Land, Water and Planning, 2022) ('*Water is Life*').

113 'Traditional Owner Nations' is the term used in the policy document and reflects the terminology used in the *TOS Act*. These policy statements are not part of the Victorian government's official policy; ibid 20.

114 *NSW Aboriginal Water Strategy* (Web Page). https://water.dpie.nsw.gov.au/plans-and-programs/aboriginal-water-program/strategy.

115 *NSW Water Strategy* (Footnote 110). See Priority 2, which is to 'Recognise First Nations/Aboriginal People's rights and values and increase access to and ownership of water for cultural and economic purposes'. It then sets out how this will be achieved, which includes by [d]evelop[ing] a state-wide Aboriginal water strategy (cl 2.2).

116 *First Nations Water Strategy* (Web Page, 4 August 2023). https://www.rdmw.qld.gov.au/water/consultations-initiatives/first-nations-water-strategy.

strategy—promoting … co-management of resources with the Ngunnawal Traditional Custodians' and '[a] Ngunnawal Water Initiative'.[117] The NT has recently released its first whole-of-territory water policy in which one of the 16 priority actions is 'enhancing Aboriginal participation in water management'.[118] A further priority action is 'understanding the Aboriginal cultural values of water'.[119] In 2022, South Australia released its *Water Security Statement 2022: Water for Sustainable Growth*.[120] One of the 10 priorities in the statement is to 'Ensure First Nations have equitable access to water resources for cultural purposes, including economic purposes'.[121] Tasmania has no overarching state water policy or First Nations water policy; however, it has commenced engagement with First Nations Tasmanians to understand their interests in and connection to water.[122]

These existing or emerging policies all reflect an increasing awareness by government of the need to recognise, protect and provide for First Nations water rights and interests and to incorporate First Nations water knowledge into their water management regimes. This is an important and positive development. However, as policies do not have the force of law, as a mechanism for facilitating a just transition for First Nations Australians, they are vulnerable to changing political priorities.

4.5 Conclusion

It is imperative for a just transition to decarbonisation that First Nations water knowledge be recognised in Australia's water law and policy, and thus contribute to sustainable water management. Equally important is the recognition, protection of and provision for First Nations peoples'

[117] ACT Government, *Fact Sheet: ACT Water Governance Review* (n.d.). https://www.environment.act.gov.au/water/office-of-water.

[118] Office of Water Security, *Territory Water Plan* (June 2023) 31. https://watersecurity.nt.gov.au/territory-water-plan.

[119] Ibid., 32.

[120] (Department of Environment and Water, 2022). https://www.environment.sa.gov.au/topics/water/water-security/water-security-statement.

[121] Ibid., 49.

[122] Department of Natural Resource and Environment Tasmania, *Tasmanian Aboriginal Freshwater Interests: What are we Learning?* (December 2022). https://nre.tas.gov.au/water/water-legislation-policies-and-strategies/aboriginal-freshwater-interests.

water access and use, for them to be able to maintain their spiritual and cultural connection to Country. As demonstrated above, there has been some movement in recent years, with the policy landscape seeing a significant shift in that regard. However, policy is not legally enforceable. Nonetheless recent developments are a step in the right direction as they are an explicit recognition of the vital role of First Nations knowledge, and their rights and obligations to water, that are necessary for a just transition to decarbonisation. It is just as important to ensure that there is also a legislative basis for this because legislation can create 'an official space for an Indigenous voice'.[123] In that regard, Australian water law still has a long way to go.

[123] William Nikolakis et al. (Footnote 7) 180.

CHAPTER 5

Utilising Free Trade Agreements to Protect Nature and Heritage

Abstract This chapter revolves around the question of whether free trade agreements (FTAs) should/could be utilised in redressing the harm inflicted by climate change upon the natural environment—notably biodiversity—and its concomitant effect on the intertwined cultural diversity, heritage and vulnerable Indigenous peoples, and thereby have the potential to contribute to a just transition to a low emission and resilient economy. Climate change has irreparably impacted the natural environment, including the ecosystems of areas essential for Indigenous peoples' wellbeing, some of which are areas inscribed on the World Heritage List—pledged and designated to be preserved for the benefit of all humankind. As climate change and biodiversity crises worsen there is the urgency that they be tackled on various fronts, including trade due to its vast prominence in the global economy. Given that FTAs are being increasingly negotiated throughout the world, the question of their utility vis-à-vis the protection of nature is becoming ever more significant. By taking a biocultural heritage approach, it is argued that in transitioning away from a fossil fuel economy, FTAs have the potential to play an important role in the context of international trade, climate change, biodiversity and heritage, and that this transition cannot be seen as just without an adequate place being given to Indigenous peoples.

Keywords Biodiversity · Climate change · FTAs · Heritage · Indigenous peoples

D. Kraal et al., *A Just Transition to Decarbonisation*, Just Transitions, https://doi.org/10.1007/978-3-031-49137-5_5

5.1 Introduction and an Overview

This chapter concerns the just transition approach to decarbonisation with a focus on the interplay between international trade, climate change, biodiversity and heritage, and the implications of this interplay for Indigenous peoples. The relationship between international trade, climate change, biodiversity, heritage, and a just transition to decarbonisation is complex and multifaceted. The impact of trade, especially trade in fossil fuels, including coal, on climate change and the natural environment—including ecosystems housed in heritage sites—has been well documented.[1] The ever-expanding volume of world trade causes rise in greenhouse gas emissions (GHG emissions) (e.g. through the types of goods being traded, their production methods and modes of transport),[2] which in turn impacts upon global warming and climate change. Trade, coal mining and climate change impacts upon biodiversity and heritage are human-caused and alarming vis-à-vis the levels of their gravity.[3] Trade

[1] International Energy Agency [IEA] Reports: *Coal 2020: Trade*. https://www.iea.org; Intergovernmental Panel on Climate Change [IPCC] Reports: *Climate Change Widespread, Rapid, and Intensifying—IPCC* (9 August 2021, Geneva), https://www.ipcc.ch/2021/08/09/ar6/; *IPCC AR6 Synthesis Report: Climate Change 2023* (13–19 March 2023, Geneva), https://www.ipcc.ch/report/sixth-assessment-report-cycle/; Victoria McGinnes and Murray Raff, 'Coal and Climate Change: A Study of Contemporary Climate Litigation in Australia' (2020) 37 *Environmental and Planning Law Journal* 87; UN News, Climate and Environment: 'Drop Coal or Climate Change will "Wreak Havoc" across Australian Economy' (6 September 2021). https://www.un.org/news; BBC News Round, 'COP26: Why Is Coal Such a Big Issue in Climate Change Talks?' (16 November 2021), https://www.bbc.com/news.

[2] World Trade Organisation [WTO], 'The Impact of Trade Opening on Climate Change', https://www.wto.org/english/tratomp_e/; OECD, 'Carbon Dioxide Emissions Embodied in International Trade'. https://www.oecd.org; WTO, *World Trade Report 2022: Climate Change and International Trade* (WTO, *Report 2022*). https://www.wto.org; Paul Brenton and Vicky Chenutai, *The Trade and Climate Change Nexus: The Urgency and Opportunities for Developing Countries* (The World Bank, 2021). https://www.worldbank.org; Xiao-Tong Wang et al., 'Trade-Linked Shipping CO_2 Emissions' (2021) 11 *Nature Climate Change* 945.

[3] Livia Cabernard and Stephan Pfister, 'Hotspots of Mining-Related Biodiversity Loss in Global Supply Chains and the Potential for Reduction Through Renewable Electricity' (2022) 56(22) *Environmental Science & Technology* 16357; Carla Howarth, 'Australia's World Heritage Sites are Facing Unprecedented Challenges from Climate Change' (28 July 2023). https://www.csiro.org.

itself has also been affected by climate change,[4] and it has been seen as the potential contributor to the fight against climate change.[5]

Intergovernmental Panel on Climate Change Reports (IPCC Reports)[6] indicate that planet Earth is facing an unprecedented climate emergency.[7] One aspect of this emergency has been manifested through loss of both biodiversity and cultural diversity, and the interrelated biocultural heritage. To reiterate, these losses are human-caused and unparalleled in human history.[8]

Climate change and biodiversity loss are intertwined.[9] The interaction between climate change and biodiversity loss, and the urgency of the need to consider these pressing issues as two sides of the same coin, has been recognised, including at the recently held 27th Session of the Conference of the Parties to the United Nations *Framework Convention on Climate Change* (‘*UNFCCC*’),[10] commonly referred to as COP27, held in Sharm El-Sheikh, Egypt, from 6 to 18 November 2022,[11] and at the United Nations *Convention on Biological Diversity's* (‘*CBD's*’ or ‘*Biodiversity Convention's*’)[12] Fifteenth Conference of the Parties, usually

[4] WTO, *Report 2022* (Footnote 2).

[5] Charlotte Janssens et al., ‘International Trade is a Key Component of Climate Change Adaptation’ (2021) 11 *Nature Climate Change* 915.

[6] IPCC Reports are available at, https://www.ipcc.ch.

[7] *IPCC AR6 Synthesis Report* (Footnote 1).

[8] Victoria Reyes-Garcia et al., ‘Biocultural Vulnerability Exposes Threats of Culturally Important Species’ (2023) 120(2) *Anthropology* e2217303120.

[9] Miguel Mahecha et al., ‘Biodiversity Loss and Climate Extremes—Study the Feedbacks: Enough of Silos: Develop a Joint Scientific Agenda to Understand the Intertwined Global Crises of the Earth System’ (*Nature*, 29 November 2022). https://www.nature.com; Georgina Gustin, ‘New Report: Climate Change and Biodiversity Loss Must be Tackled Together and Not Separately’ (Inside Climate News, 11 June 2021). https://www.insideclimatenews.org; IPBES-IPCC Co-Sponsored Workshop, ‘Biodiversity and Climate Change’—Workshop Report (2020), IPBES_IPCC_WR_12_2020.pdf; UN Climate Change, ‘New International Biodiversity Agreement Strengthens Climate Action’ (19 December 2022). https://www.un.org.

[10] *United Nations Framework Convention on Climate Change*, 1771 UNTS 107, opened for signature 9 May 1992 (entered into force 21 March 1994) [‘*UNFCCC*’].

[11] UN, ‘COP27: Protecting Biodiversity is Protecting the *Paris Agreement*’ (UN News, 16 November 2022). https://news.un.org/en/story/2022/11/1130677.

[12] *Convention on Biological Diversity*, 1760 UNTS 79, open for signature 5 June 1992 (entered into force 29 December 1993) [‘*CBD*’] (as of 4 September 2023, there are 168 Signatories and 196 State Parties).

mentioned in its abbreviated form, COP15, held in Montreal, Canada, from 7 to 19 December 2022.[13] Given the exigent nature of climate change[14] and of biodiversity loss,[15] together with the associated heritage loss, various ways and approaches have been utilised to tackle these emergencies.[16] There is mounting pressure for the ever-increasing number of free trade agreements (FTAs), concluded by states on a bilateral or multilateral basis,[17] to address biodiversity issues and include non-trade provisions.[18] While the international community is transitioning away from a fossil fuel-based economies, there is the need for the transition to be just and to address the needs of those vulnerable and marginalised, including Indigenous peoples.

There is a wealth of the literature on climate change,[19] and on biodiversity,[20] and an ever-growing body of literature on a just transition

[13] Jeffrey Qi and Anika Tertan, 'From Sharm el-Sheikh to Montreal: Seizing the Moment for the Biodiversity—Climate Nexus at COP15' (IISD, 9 December 2022). https://www.iisd.org.

[14] IPCC, 'Climate Change Widespread, Rapid and Intensifying—IPCC' (9 August 2021). https://ipcc.ch/2021/08/09/.

[15] Forest Isbell et al., 'Expert Perspectives on Global Biodiversity Loss and its Drivers and Impacts on People' (*Frontiers in Ecology and the Environment*, 2023). https://esajournals.onlinelibrary.wiley.com.

[16] Christopher Dent, 'Trade, Climate and Energy: A New Study on Climate Action through Free Trade Agreements' (2021) 14 *Energies* 4363.

[17] WTO, 'Regional Trade Agreements'. https://www.wto.org (as of 1 August 2023, 360 RTAs were in force). The abbreviations 'RTAs' (regional trade agreements) and 'FTAs' (free trade agreements) are used interchangeably in this chapter.

[18] Emily Benson and Lexie Judd, 'Trade Laws of Nature: Biodiversity Provisions and the AfCFTA' (22 November 2021). https://www.csis.org; Congressional Research Service, 'Environmental Provisions in Free Trade Agreement (FTAs)' (3 February 2021). https://www.everycrsreport.com; EIAS, 'The European Commission's New Approach to Trade Agreements: The Impact on EU-ASEAN FTAs' (30 September 2022). https://eias.org/publications/.

[19] Lea Berrang-Ford et al., 'A Systematic Global Stocktake of Evidence on Human Adaptation to Climate Change' (2021) 11 *Nature Climate Change 989*; Stefee Bohn and Sian Sullivan (eds), *Negotiating Climate Change in Crisis* (Open Book Publishers, 2021); Lucy Holmes-McHugh et al., 'Risk? Crisis? Emergency? Implications of the New Climate Emergency Framing for Governance and Policy' (2021) 1296 *WIREs Wiley Interdisciplinary Reviews* e736; Hans Baer and Merrill Singer, *The Anthropology of Climate Change: An Integrated Critical Perspective* (Routledge, 2nd ed, 2018).

[20] James Albert et al., 'Scientists' Warning to Humanity on the Freshwater Biodiversity Crisis' (2021) 50 *Ambio* 85; MJ Sanders, 'A California Court Decision—Bees are

to decarbonisation[21]—namely, the transition to a low-carbon economy, which encompasses various forms of justice.[22] Despite this, little has been written on the subject of utilising FTAs to address the implications of extracting of and trade in fossil fuels, predominantly coal, for climate change, biodiversity and heritage, in the Australian context; specifically effects of these phenomena on vulnerable communities, particularly Australia's First Nations peoples, and the importance of involving these peoples in the relevant decarbonisation processes.[23] This chapter aims to contribute to scholarship in this area.

Fish—Shines New Light on the Biodiversity Crisis' (2022–2023) 54 *Trends* 5; Jessica Stubenrauch and Beatrice Garske, 'Forest Protection in the EU's Renewable Energy Directive and Nature Conservation Legislation in Light of the Climate and Biodiversity Crisis' (2023) 153 *Forest Policy and Economics* 102996; Hilary Whitehouse, 'The Extinction Crisis' in *The Routledge Handbook of Global Development* (Routledge, 2022) 22.

[21] Raphael Heffron, 'What is the "Just Transition"?' in *Achieving a Just Transition to a Low Carbon Economy* (Palgrave MacMillan, 2021); Giuseppe Feola et al., 'Collective Memories, Place-Framing and the Politics of Imaginary Futures in Sustainability Transitions and Transformation' (2023) 138 *Geoforum* 103668; George Boss et al., 'Politics, Voice and Just Transition: Who Has a Say in Climate Change Decision Making, and Who Does Not' (2023) *Global Social Challenges Journal* 1; Raphael Heffron, 'Repurposing for the Just Transition: Energy Companies Need to Future-Proof Their Structure and Strategy' (2023) 16(3) *The Journal of World Energy Law and Business* 302.

[22] Raphael Heffron, 'The Just Transition to a Low-Carbon Economy' (2018) 8(4) *Renewable Energy Law and Policy Review* 39.

[23] Mareese Terare and Margot Rawsthorne, 'Country Is Tarning to Me: Worldview, Health and Well-Being Amongst Australian First Nations People' (2020) 50(3) *The British Journal of Social Work* 944 (explaining that First Nations Australians include Aboriginal and Torres Strait Islanders Peoples and Indigenous Australians); AIATSIS, 'Australia's First Peoples'. https://aiatsis.gov.au (observing that the term 'Indigenous Australian' is very general and that the two very distinct cultural groups—Aboriginal and Torres Strait Islander Peoples—are umbrella terms, within which sits a large array of different nations, each with their own culture, language, beliefs and practices, and advising that there is a growing preference for the term 'First Nations Australians' as a more encompassing term, because while it also is generic, it acknowledges the diversity of Australia's First Peoples); National Museum Australia, 'Evidence of First Peoples'. https://www.nma.gov.au (noting that Australian First Nations Peoples, or Indigenous Australians, are known to have occupied mainland Australia for at least 65,000 years). See also, UN Department of Economic and Social Affairs: Indigenous Peoples, 'Indigenous Peoples at the UN'. https://www.un.org (the term 'Indigenous Peoples' is commonly used in international law). In this chapter, all these terms are used interchangeably.

Australia is among the world leaders in coal production and coal export.[24] At the same time, it is a region with among the richest biodiversity on the planet.[25] Some of its ecosystems are housed in outstanding World Heritage sites whose preservation is in the interest of Australia's First Nations peoples who have strong connections to the Australian environment; other local communities and ultimately all humankind, including generations yet to come.[26] Australia is also party to a number of FTAs (18 FTAs were in force for Australia at the time of this writing),[27] including the *Comprehensive and Progressive Agreement for the Transpacific Partnership* ('*CPTPP*')[28]—the so-called 'mega' FTA (or 'new generation agreement' as some scholars prefer to call it).[29] The *CPTPP* was signed on 8 March 2018 in Santiago, Chile, and entered into force on 30 December 2018 for its 11 signatory States: Australia, Canada, Japan, Mexico, New Zealand, Singapore, Vietnam, Peru, Malaysia, Chile

[24] Report Linker, 'Coal Industry 2023'. https://www.reportlinker.com; Katrina Beavan, 'Coal Exports Forecast to Smash Record With Value Set to Break $100 Billion This Financial Year' (BBC News, 4 April 2022). https://www.bbc.com/news (noting that about 70 percent of coal mined in Australia is exported; in 2019–2020 Australia was the world's largest exporter of thermal coal and second largest exporter of metallurgical coal; in April 2022 it was reported that coal was the second commodity to reach $100 billion annual export figure, only behind iron ore).

[25] Australian Government, Department of Climate Change, Energy, the Environment and Water [DCCEEW], 'Protecting Biodiversity'. https://www.dcceew.gov.au (together with 16 other countries, Australia is a megadiverse country, one of the most biologically rich countries in the world, with 9.6 percent of all species known to exist, large percentage of which are endemic, or unique to Australia).

[26] Australia has a number of sites of outstanding universal value inscribed on the UNESCO World Heritage List, some of which provide home for large ecosystems such as those in the Great Barrier Reef, the Kakadu National Park, the Wet Tropics of Queensland and in the Tasmanian Wilderness. See, eg, UNESCO World Heritage Centre—World Heritage List. https://whc.unesco.org/en/list.

[27] Australian Department of Foreign Affairs and Trade [DFAT], 'Free Trade Agreements in Force'. https://www.dfat.gov.au (last visited September 2023).

[28] *Comprehensive and Progressive Agreement for the Transpacific Partnership*, opened for signature 8 March 2018 (entered into force 30 December 2018) ['*CPTPP*'].

[29] Emily Hush, 'Where No Man Has Gone Before: The Future of Sustainable Development in the Comprehensive Economic and Trade Agreement and New Generation Free Trade Agreements' (2018) 43(1) *Columbia Journal of Environmental Law* 93.

and Brunei Darussalam.[30] On 16 July 2023, the UK also signed its *Accession Protocol* to join the *CPTPP*.[31] This agreement is predicted to be a leader among trade deals[32] and is likely to influence other international trade agreements. Further, the subject of trade is envisioned to occupy a considerable time and space in the upcoming United Nations Climate Change Conference, to be held in Dubai, United Arab Emirates, from 30 November to 12 December 2023 (COP28).[33] For the first time in the history of these high-level international climate talks, issues of trade will be considered,[34] adding to the significance of discussing trade and the environment in the context of climate change and a just transition to decarbonisation. This makes the overall topic of this chapter timely and relevant.

By drawing on various empirical studies and other scholarly writing, assorted reports, relevant provisions of the applicable legal instruments and other literature, and by focussing on the *CPTPP* and the Australian jurisdiction and using the Great Barrier Reef—a World Heritage site—as an example, this chapter examines how the issues of climate change and the protection of biodiversity and heritage have been dealt with by mega FTAs or new generation agreements. The chapter takes a biocultural heritage approach emphasising Indigenous People's role in the context of a just transition to decarbonisation vis-à-vis biodiversity and heritage. Biocultural heritage considerations are augmented by sustainability considerations that place intergenerational justice at the centre. The chapter argues that in transitioning away from a fossil fuel economy, FTAs have the potential to play an important role in the context of international trade, climate change, biodiversity and heritage, and that this transition cannot be seen as just without a prominent place being given to Indigenous peoples.

[30] DFAT, 'About the *Comprehensive and Progressive Agreement for Trans-Pacific Partnership* (CPTPP)'. https://www.dfat.gov.au.

[31] DFAT, 'United Kingdom Formal Accession to the *Comprehensive and Progressive Agreement for Trans-Pacific Partnership* (CPTPP)'. https://www.dfat.gov.au.

[32] Raj Bhala, 'Trans-Pacific Partnership or Trampling Poor Partners? A Tentative Critical Review' (2014) 11 *Manchester Journal of International Economic Law* 2, 4, 5.

[33] UNCTAD, 'Trade's Role in Climate Action in the Spotlight at COP28' (UNCTAD Climate Change News, 20 July 2023). https://www.unctad.org.

[34] UN Climate Change website, 'COP28'. https://unfccc.int/cop28.

The remainder of this chapter first defines key concepts and highlights the intersection between the factors examined and then, after overviewing the approach to a just transition to decarbonisation and the biocultural heritage approach, it canvasses the potential role of FTAs in addressing the climate change–biodiversity crisis and underlines the importance of Indigenous peoples' involvement in the processes relevant to this just transition. This is followed by a sketching out of the implications of climate change and trade for biodiversity and heritage sites in Australia, and a survey of the relevant responses provided by the *CPTPP*. The final part of the chapter suggests ways forward, identifies areas that warrant further research and offers some concluding remarks.

5.2 Key Concepts

5.2.1 Identifying Key Concepts

This chapter rests on several major concepts. The first main notion encompassed by this chapter is biological diversity or biodiversity. The *Biodiversity Convention* defines biodiversity as 'the variability among living organisms from all sources, including, *inter alia*, terrestrial marine and other aquatic systems and the ecological complexes of which they are part; this includes diversity within species, between species, and of ecosystems'.[35] Biodiversity is important to humanity for multiple reasons, including its role in the global economy, its functions in culture and identity and its links to human physical health.[36] Research has shown that biodiversity is also the key to ecosystem health, as it serves as an absorptive barrier, providing protection from environmental shocks and stresses.[37]

Biodiversity is considered part of humanity's heritage. Heritage is a broad concept that includes cultural heritage—both tangible and intangible traits of a group or society that are inherited from past generations and supposed to be handed over to future generations—and natural heritage, which encompasses the countryside, natural environment and biodiversity. Some heritage is a combination of cultural heritage and

[35] *CBD*, art 2.

[36] Carmen Ang, 'WWF: These are the Biggest Threats to the Earth's Biodiversity' (*Nature and Biodiversity*, 16 Novembewr 2020). https://www.weforum.org.

[37] S. Stolton et al., *Natural Security: Protected Areas and Hazard Mitigation* (Gland: WWF International, 2008).

natural heritage. The *Convention concerning the Protection of the World Cultural and Natural Heritage* (*World Heritage Convention*),[38] recognises some of the world's properties specifically for their outstanding universal biodiversity values, protecting many important ecosystems and areas of high biodiversity on the planet.[39]

Biodiversity is related to cultural heritage in several ways, including being part of natural heritage. Natural heritage is an important part of a society's heritage, and it includes flora and fauna, as well as geological elements, scientifically known as geodiversity.[40] Biodiversity is related to cultural heritage through cultural value; namely, biodiversity has cultural value to humans, such as for spiritual or religious reasons.[41] Therefore, biodiversity is an important component of cultural heritage, and its conservation and sustainable use are important considerations in heritage preservation efforts.

Concomitantly, biodiversity is interrelated to cultural diversity. There is no uniform definition of the concept of cultural diversity as there is no universal definition of a culture.[42] Its meaning depends on the context, but broadly it can be defined as differences in race, ethnicity, language, nationality or religion among various groups within a community, an organisation or a nation.[43] It is embodied in the uniqueness and plurality

[38] *Convention concerning the Protection of the World Cultural and Natural Heritage*, opened for signature 16 November 1972, 1037 UNTS 151 (entered into force 17 December 1975) [*World Heritage Convention*].

[39] Ibid art 2.

[40] Ibid (natural heritage is a term used by UNESCO to describe 'natural features consisting of physical and biological formations or groups of such formations, which are of outstanding universal value from the aesthetic or scientific point of view; geological and physiographical formations and precisely delineated areas which constitute the habitat of threatened species of animals and plants of outstanding universal value from the point of view of science or conservation; natural sites or precisely delineated natural areas of outstanding universal value from the point of view of science, conservation or natural beauty').

[41] In contrast, the intrinsic value of biodiversity refers to its inherent worth, independent of any benefit to humans.

[42] C. Geertz, *The Interpretations of Culture* (Basic Books, New York, 1973) (the notion of culture can be defined in multiple ways).

[43] Law Insider, Definition of Cultural diversity. https://www.lawinsider.com/dictionary/cultural-diversity.

of the identities of the groups and societies making up humankind,[44] and it is important for humankind just as biodiversity is for nature.[45]

Many cultural landscapes safeguard important biodiversity values, often based on interlinkages between cultural and biological diversity.[46] The concept of biocultural diversity recognises the nexuses between biological diversity and cultural diversity, considering them as interconnected aspects of heritage.[47] Areas of high cultural diversity are often areas of high biological diversity.[48] Different cultures interact with natural environment in different ways and forge different relationship with their local environments.[49] Research reveals that the degree to which biodiversity is linked to its cultures is only beginning to be understood and that ironically, it is precisely as our knowledge of this linkage is advancing that these cultures are receding in various forms such as endangered species, threatened habitats, dying languages and knowledge bases, and that any hope for saving biological diversity is heavily reliant on an interrelated effort to respect and preserve cultural diversity.[50]

The correlated paradigm of biocultural heritage has been described as 'a complex system of interdependent parts centred on the relationship between Indigenous peoples and their natural environment'.[51] The United Nations Educational, Scientific and Cultural Organisation

[44] *Universal Declaration on Cultural Diversity*, adopted by the 31st Session of the General Conference of UNESCO, Paris, 2 November 2001. https://en.unesco.org/about-us/legal-affairs/unesco-universal-declaration-cultu....

[45] Ibid.

[46] Mauro Agnoletti and Ian Rotherham, 'Landscape and Biocultural Diversity' (2015) *24 Biodiversity and Conservation* 3155.

[47] Peter Bridgewater and Suraj Upaghaya, 'What is Biocultural Diversity, and Why Does It Matter?' (*The Conversation*, 16 February 2022). https://www.theconversation.com.

[48] The biocultural heritage has been promoted by UNESCO and the Secretariat of the CBD through their Joint Programme on Biological and Cultural Diversity since 2010. See, UNESCO, World Heritage Centre, 'World Heritage and Biodiversity'. https://www.unesco.org.

[49] K Milton, 'Nature and the Environment in Indigenous and Traditional Cultures' in DE Cooper and JA Palmer (eds), *Spirit of the Environment: Religion, Value and Environmental Concern* (Routledge, 1998).

[50] Jules Pretty et al., 'The Intersections of Biological Diversity and Cultural Diversity: Towards Integration' (2010) 7(2) *Conservation and Society* 100.

[51] International Institute for Environment and Development (IIED), 'Biocultural Heritage'. https://www.iied.org.

(UNESCO) defines biocultural heritage as 'living organisms or habitats whose present features are due to cultural action in time and place'.[52] Biocultural heritage 'encompasses the natural and cultural components of human and environment interaction, including knowledge, practices and innovation'.[53] Biocultural heritage is closely linked to First Nations peoples' cultural identity and religious beliefs.[54] This type of heritage includes biological resources and generations–long (sometimes millennia–long) traditions, practices and knowledge for adaptation to environmental change and sustainable use of biodiversity.[55] While important for food and health security generally, this collectively held heritage, sustains local economies and is vital for survival of First Nations peoples who depend on natural resources and are vulnerable to climate change risks.[56]

The phenomenon of climate change is defined as 'a change of climate that is attributed directly or indirectly to human activity that alters the composition of the global atmosphere and that is in addition to natural climate variability observed over comparable time periods'.[57] Climate change poses a significant threat to biodiversity, biocultural heritage and biocultural diversity.

5.2.2 *Interplay of Key Concepts and Major Issues*

5.2.2.1 *Biodiversity Loss*

Impact of Climate Change on Biodiversity Loss

Biodiversity loss is one of the major repercussions of climate change. Climate change has altered marine, terrestrial and freshwater ecosystems around the world, causing the loss of local species, increased spread of diseases and other negative effects. Climate change is currently considered to be the second biggest cause of biodiversity loss in the ocean

[52] Ana Persic and Gary Martin, *Links between Biological and Cultural Diversity: Concepts, Methods and Experiences* (*Report of the International Workshop*, UNESCO, Paris 2007).

[53] Krystyna Swiderska, 'What is Biocultural Heritage?' (March 2017). https://www.researchgate.net/.

[54] Ibid.

[55] Persic and Martin (Footnote 52).

[56] Swiderska (Footnote 53).

[57] *UNFCCC* art 1(2).

and the fourth biggest cause on land,[58] and believed likely to play a greater role in these processes in the future.[59] If the current trends in biodiversity loss continue, it is feared that one million animal and plant species will become threatened with extinction,[60] which has never happened at any other point in human history. Through the effects of climate change, species are being forced to move out of areas in which they have evolved for millions of years as those areas become uninhabitable for them.[61] Together with GHG emissions, the underlying cause of climate change and its impacts on biodiversity is humans' unsustainable use of resources.[62] Climate change is fundamentally changing the ecosystems in which many species live, threatening their continued health and survival.[63] Corals, birds, amphibians and trees are some examples of species affected by climate change.[64]

The consequences of biodiversity loss due to climate change are grave and wide-ranging. They include loss of ecosystem services; increased risk of diseases; reduced resilience to climate change and species extinction. It is also believed that if current trends in biodiversity loss continue, not only that many more species will be threatened with extinction, but that this will have cascading effects on entire ecosystems.[65] According to the International Union for Conservation of Nature's Red List of Threatened Species List (IUCN's Red List), more than 10,000 species[66] of the

[58] Holly Chetan-Welsh and Lisa Hendry, 'How are Climate Change and Biodiversity Loss Linked?' (Natural History Museum). https://www.nhm.ac.uk.

[59] Ibid.

[60] Helen Briggs, 'Biodiversity Loss Risks "Ecological Meltdown" Scientists' (BBC News, 10 October 2021). https://www.bbc.com/news.

[61] Craig Welch, 'Half of all Species are on the Move—and We're Feeling It' (*National Geographic*, 27 April 2017). https://www.australiangeographic.com.au.

[62] KR Shivanna, 'Climate Change and Its Impact on Biodiversity and Human Welfare' (2022) 88(2) *Procindian Nationall Science Academy* 160.

[63] The Royal Society, 'How Does Climate Change Affect Biodiversity?' https://www.theroyalsociety.org; Australian Academy of Science, 'What are the Impacts of Climate Change?' https://www.science.org.au/.

[64] Katie Wood and Lauretta Burke, 'What Must be Done to Protect Coral Reefs, Experts Explain' (World Economic Forum, 12 January 2022). https://www.weforum.org.

[65] Stephanie D'Agata, 'Ecosystems Services at Risk' (2022) 12 *Nature Climate Change* 13.

[66] International Union for Conservation of Nature (IUCN) Red List of Threatened Species. https://www.iucn.org.

world's 42,100 threatened species listed therein are currently affected by climate change, increasing the likelihood of their extinction.[67] Reportedly, many species are already extinct due to global warming: 882 plants and animals no longer exist.[68] Some biologists warn that 35 per cent of animals and plants could become extinct in the wild by 2050 as a result of global climate change.[69]

To prevent further biodiversity losses it is important to take action to mitigate the effects of climate change on biodiversity, first and foremost by reducing GHG emissions, which are the primary cause of climate change and predominantly stem from fossil fuel extraction and burning, most notably coal.[70] At the COP15, the governments of close to 200 states adopted the *Global Biodiversity Framework* ('*GBF*') as a step towards the United Nations' (UN's) 2050 vision for biodiversity of 'living in harmony with nature'.[71] One of the *GBF*'s principal goals for 2030 is to reduce the risks to biodiversity. COP15's target is to protect 30 per cent of Earth's lands, oceans, coastal areas and inland waters—the so-called, 30 × 30 target—(seen as the biodiversity counterpart of the *Paris Agreement*'s[72] goal to limit global warming to 1.5 °C above pre-industrial levels). This target is seen as overly enthusiastic, however, as it requires a near doubling of land area and a tripling of marine areas currently under protection, which has not produced the desired results. Also, it has been argued that such area-based conservation could, *inter alia*, undermine the rights of Indigenous peoples and local communities.[73] While Target Three of the

[67] Holly Dublin, 'IUCN Red List of Threatened Species' (*Britannica*, 8 September 2023).

[68] United Nations, 'World is "on notice" as Major UN Report Shows One Million Species Face Extinction' (UN News, 6 May 2019). https://www.un.org; IPBES, *Global Assessment Report on Biodiversity and Ecosystem Services* (2019). https://www.ipbes.net/global.assessment.

[69] National Park Service, 'Climate Change Endangers Wildlife'. https://www.nps.gov.

[70] Laura Sonter et al., 'Mining and Biodiversity: Key Issues and Research Needs in Conservation Science' (*Proceedings of the Royal Society B: Biological Sciences*, 5 December 2018). https://www.royalsocietypublishing.org.

[71] United Nations, 'Biodiversity Commitments Builds Hope for "Living in Harmony with Nature"' (UN, 13 October 2021). https://www.un.org/news.

[72] *Paris Agreement*, adopted 12 December 2015, opened for signature 22 April 2016 (entered into force 4 November 2016) art 2(1)(a) [*Paris Agreement*].

[73] Oxfam, 'The COP15 Target 30 × 30 a Threat to the Rights of Indigenous Peoples' (Oxfam, 16 December 2022). https://www.oxfam.org.

GBF includes the rights of Indigenous peoples and local communities including over their traditional territories, which inclusion has been seen by some Indigenous peoples present at COP15 as transformational (as the non-inclusion of the rights in this goal could, in their view, result in Indigenous displacement and other human rights abuses in the name of conservation), many Indigenous activists are of the view that COP15 did not go far enough to protect and uplift Indigenous peoples.[74] Moreover, others reference procedural limitations at COP15 that prevented the full recognition of Indigenous perspectives as Indigenous Nations are not on the list of parties to the CBD, thereby having unequal status during negotiations.[75] On a positive note (given that coal power generation is a primary cause of GHG and toxic airborne emissions globally—resulting in up to three trillion dollar in damages worldwide every year),[76] this will require the mining industry to adapt to higher environmental standards and scrutiny, where mining companies are expected to face increasingly stringent permitting processes for new activities, with a rising percentage of proposed new projects likely to be rejected on biodiversity grounds (as well as on overall environmental impact assessment grounds).[77]

Impact of Illicit Trade in Wildlife on Biodiversity Loss

Illicit trade in wildlife is another major threat to many endangered species around the world. This trade is devastating many species worldwide and threatens iconic species with extinction.[78] The illicit trade in species is a multibillion-dollar business, estimated to be worth between seven to 23

[74] Olivia Amitay, 'How the UN Biodiversity Conference Impacts Indigenous Communities' (Peril and Promise, 5 April 2023). http://www.pbs.org.

[75] Ibid.

[76] Anastasia Zagoruichyk, 'Greenhouse Gas Emissions from Global Mining and Resource Extraction Result in up to £2.5tn ($3tn) in Damages Worldwide Every Year, According to New Study' (*Carbon Brief*, 6 July 2022). https://www.carbonbrief.org.

[77] CRU, 'Focus on Biodiversity Will Require Miners to Adapt' (30 January 2030). https://www.crugroup.com.

[78] Endangered Species Coalition, 'Trafficked: 10 Species Threatened by the Wildlife Trade' (18 December 2020). https://www.endangered.org/trafficked.

billion dollar per year.[79] This trade undermines efforts made by states to protect their natural resources.

Elephants, rhinoceroses, turtles, tigers and pangolins are some of the examples of species threatened by the illicit wildlife trade.[80] The *Convention on International Trade in Endangered Species of Wild Fauna and Flora* ('*CITES*') aims to ensure that international trade in specimens of wild animals and plants does not threaten their survival.[81] Despite fairly robust international and municipal regulation, including criminal sanctions,[82] the illegal trade in wildlife continues to inflict damage on biodiversity and extensive exploration of effective measures to prevent extinctions of precious species is required.

5.2.2.2 *Impact of Climate Change on Tangible and Intangible Aspects of Biocultural Heritage Sites*

Climate change is impacting both tangible and intangible aspects of biocultural heritage sites in various ways. For example, rising sea levels and increased storm activity are causing physical damage to cultural heritage structures and artifacts.[83] In turn, changes in temperature and precipitation patterns are affecting the ecosystems and biodiversity that are integral to many biocultural heritage sites.[84] Climate change is also affecting the

[79] IATA, 'Illegal Trade in Wildlife' Fact Sheet (this is 'the fourth largest illegal global trade (after drugs, counterfeiting and human trafficking) and covers a broad range of endangered species'). https://www.iata.org.

[80] WWF, 'Illegal Wildlife Trade: Investigations', KS3_WWF_IllegalWildlifeTradeInvestigations.pdf; WWF, 'Second-Biggest Direct Threat to Species after Habitat Destruction'. https://www.panda.org.

[81] *Convention on International Trade in Endangered Species of Wild Fauna and Flora*, opened for signature 3 March 1973 (entered into force 1 July 1975) [*CITES*].

[82] Ivonne Higuero, 'Understanding the Size and Shape of Illicit Trade in Wildlife' (UNCTAD Illicit Trade Forum, Geneva, Switzerland, 3 February 2020).

[83] Elena Sesana et al., 'Climate Change Impacts on Cultural Heritage: A Literature Review' (2021) 12(4) *WIREs Climate Change* e710.

[84] UNESCO World Heritage Convention, 'Climate Change and World Heritage'. https://www.unesco.org.

intangible aspects of biocultural heritage sites, such as traditional knowledge and cultural practices, which may be impacted by changes in the natural environment.[85]

Tangible Heritage Sites Impacted by Climate Change

Climate change has become one of the most significant threats to biocultural heritage sites. Examples of biocultural heritage sites that have been impacted by climate change include Rapa Nui (Easter Island), Polynesia, where the escalation of sea level and coastal erosion due to climate change has resulted in the loss of structures and archaeological deposits on the island.[86] World Heritage sites, too, have been affected by climate change, including sea level rise, coastal erosion and changes in temperature and precipitation patterns, which can impact, *inter alia*, these sites' outstanding universal value, integrity, authenticity and potential for economic and social development.[87] Other examples of heritage sites that have been impacted by climate change include Venice, Italy (a city sinking due to rising sea levels, causing damage to its historic buildings and infrastructure),[88] Machu Picchu, Peru (increased rainfall and landslides in the region are threatening the stability of the site)[89] and the Great Barrier Reef, Australia (where, as discussed below, climate change is causing ocean acidification and coral bleaching, which is damaging the Reef and threatening the survival of its marine life).[90]

[85] Gül Aktürk and Martha Lerski, 'Intangible Cultural Heritage: A Benefit to Climate-Displaced and Host Communities' (2021) 11 *Journal of Environmental Studies and Sciences* 305.

[86] Google Arts & Culture, 'A Quick Guide to Rapa Nui's Climate Crisis'. https://artsandculture.google.com/story/.

[87] UNESCO (Footnote 84).

[88] Claudio Lavanga, 'As Sea Levels Rise, Venice Fights to Stay above Waterline' (NBC News, 16 February 2020). https://www.nbcnews.com.

[89] UNESCO World Heritage Convention, 'Torrential Rains Cause Death and Damage at Peruvian World Heritage Sites' (UNESCO News, 28 January 2010). https://www.unesco.org.

[90] DCCEEW, 'World Heritage Committee Decision on the Great Barrier Reef' (News, 14 September 2023). https://www.dcceew.gov.au.

Intangible Aspects of Biocultural Heritage Sites Impacted by Climate Change

Climate change is impacting intangible aspects of biocultural heritage sites in complex and interconnected ways. Traditional knowledge and cultural practices are integral to many biocultural heritage sites, yet the natural environment in such places is being affected.[91] Traditional knowledge and practices can relate to agriculture and food systems, water management, natural resource management, spiritual and cultural practices and art and crafts.[92] Biocultural heritage sites often have significant social and spiritual significance for Indigenous peoples, and climate change impacts these connections and relationships.[93] Climate change impacts on biocultural heritage sites can also have economic and livelihood effects, especially where Indigenous peoples and other local communities rely on tourism or other economic or sustenance activities related to the site.[94]

5.2.2.3 Effect of Biodiversity Loss and Climate Change on Indigenous Peoples

Indigenous peoples are defined by the United Nations Permanent Forum on Indigenous Issues as having:

> historical continuity or association with a given region or part of a given region prior to colonisation or annexation; identify themselves as indigenous and be accepted as members by their community; have strong links to territories, surrounding natural resources and ecosystems; maintain at least in part, distinct social, economic and political systems; maintain, at least in part, distinct languages, cultures, beliefs and knowledge systems; are resolved to maintain and further develop their identity and distinct

[91] Guy Jackson, 'How Climate Change is Affecting Cultural Heritage' (Lund University, 8 September 2021). https://lunduniversity.lu.se/article.

[92] Okologial Kutalokazpont, 'Study: Social Justice for Traditional Knowledge Holders Will Help Conserve Europe's Nature' (28 August 2023). https://www.phys.org.

[93] UNESCO Climate Change, 'Culture and Climate Change'. https://www.unesco.org.

[94] Claire Cave, 'Climate Change and Heritage: An Introduction' in *Fifty Years of the World Heritage Convention: Shared Responsibility—Conflict and Reconciliation* (Heritage Studies Book Series, 2022) 215.

> languages, cultures, beliefs and knowledge systems; are resolved to maintain and further develop their identity and distinct social, economic, cultural and political institutions as distinct peoples and communities; and often form non-dominant sectors of society.[95]

It is believed that at least one quarter of the world's land area is owned, managed, used and occupied by Indigenous peoples.[96] Although nature in these areas is degrading at a slower rate than in others, Indigenous peoples have been directly and heavily impacted by climate change and biodiversity loss.[97] First Nations peoples are already generally marginalised segments of a society both politically and economically.[98] Biodiversity loss and climate change deepen their vulnerability and pose additional risks for their heritage as these phenomena threaten cultural dimensions of lives and livelihoods that include the material and lived aspects of culture, identity, community cohesion and sense of place.[99] Moreover, although they possess wealth of traditional knowledge and practices needed in the preservation of nature, Indigenous peoples' role is poorly recognised in most biodiversity strategies and targets.[100]

5.3 Just Transition to Decarbonisation and the Biocultural Heritage Approach

The concept of decarbonisation encompasses reducing economies' dependence on fossil fuels in response to climate change and other sustainability challenges.[101] As noted, the *Paris Agreement* has the long-term goal of

95 UNEP, 'Indigenous Peoples and the Nature They Protect' (8 June 2020). https://www.unep.org.

96 Ibid.

97 Ibid.

98 UN Department of Economic and Social Affairs—Indigenous Peoples, 'Climate Change'. https://www.un.org/development.

99 WN Adger et al., 'Cultural Dimensions of Climate Change Impacts and Adaptation' (2013) 3 *Nature Climate Change* 112.

100 Krystyna Swiderska, 'Protecting Indigenous Cultures is Crucial for Saving the World's Biodiversity' (*The Conversation,* 14 February 2020). https://www.theconversation.com.

101 Ryan Hanna and David Victor, 'Marking the Decarbonisation Revolutions' (2021) 6 *Nature Energy* 568.

keeping the increase in global average temperature to well below 2 °C above pre-industrial levels and the aim to limit the increase to 1.5 °C.[102] There is an associated urgent need to reduce GHG emissions by 2030 so that emissions peak as soon as possible, and to transition to net-zero emissions by 2050.[103] This transition to a zero-carbon society must be just, addressing the needs of various stakeholders, including vulnerable communities, such as Indigenous peoples.

The concept of a just transition originated in the US in the 1970s where it was used in a movement by trade unions to protect workers affected by new water and air pollution regulation under environment statutes.[104] The concept has evolved since then. Now it comprises an integrated, whole-system perspective on justice, in various forms of justice, such as procedural, distributive, recognition and restorative justice.[105] In other words, a just transition approach to decarbonisation involves a set of principles, processes and practices that aim to ensure that the transition to a post-carbon economy is just and equitable, and that all communities, workers and countries are lifted up.[106] Therefore, although the original concept of a just transition was primarily concerned with matters related to job security,[107] it has evolved to pertain to issues well beyond this.[108] The approach differs from a traditional transition to decarbonisation in that it is a whole-system approach that encompasses multiple social, economic and environment-related issues and the associated forms of justice.[109] Transitioning away from coal, which is the world's most dominant and most carbon-intensive source of energy, is

102 *Paris Agreement* art 2(1)(a).

103 Nathan Cooper and Amy White, 'IPCC Report: Urgent Climate Action Needed to Halve Emissions by 2030' (World Economic Forum, 6 April 2022). https://www.weforum.org.

104 UNDP, Climate Promise, 'What is Just Transition? And Why Is It Important?' (3 November 2022). https://www.undp.org; Autumn Spanne, 'Just Transition: History, Principles, and Examples' (Treehugger, 30 June 2021). https://www.treehugger.com.

105 Simone Abrams et al., 'Just Transition: A Whole-Systems Approach to Decarbonisation' (2022) 22(8) *Climate Policy*; Vilja Johansson, 'Just Transition as an Evolving Concept in International Climate Law' (2023) 35 *Journal of Environmental Law* 229.

106 Spanne (Footnote 104).

107 UNDP (Footnote 104).

108 Abrams et al. (Footnote 105).

109 Ibid.

crucial to ensuring a clean energy future and achieving the United Nations Sustainable Development Goals (SDGs) and the above-mentioned *Paris Agreement* target.[110]

As its name suggests, the biocultural heritage approach is a holistic framework that concerns the interdependence of cultural and biological diversity.[111] This approach emphasises the importance of preserving cultural heritage and traditional knowledge, as well as protecting biodiversity and ecosystems. It puts vulnerable communities, notably Indigenous Peoples, at the centre of the processes of transitioning away from coal. The biocultural heritage approach addresses issues such as climate change, which has significant impacts on human lives and livelihoods,[112] biocultural diversity loss and how FTAs operate.

5.4 The Role of Free Trade Agreements in Preventing Biodiversity and Biocultural Losses

FTAs can play a role in preventing or reducing biodiversity loss by incorporating provisions and measures that promote sustainable trade practices and conservation efforts. FTAs can include specific, carefully drafted, environmental provisions that address biodiversity conservation and sustainable resource management and biocultural heritage. These provisions may require parties to uphold environmental standards, protect endangered species and combat illegal wildlife trade. FTAs can also encourage sustainable trade practices that minimise the negative impacts on biodiversity, such as promoting sustainable sourcing of natural resources, promoting sustainable agriculture and fisheries practices and supporting the use of sustainable certification schemes. Other measures that can be employed through FTAs are collaboration between states to address biodiversity loss and capacity building, integration with international legal frameworks relevant to biodiversity conservation, such as the above-mentioned

[110] *Paris Agreement*, art 2(1)(a).

[111] Justine Massey, 'Climate Change, Culture and Cultural Rights' (Memo in preparation for the Report by the UN Special Rapporteur in the field of cultural rights, Karima Bennoune, 15 May 2020). https://www.ohchr.org.

[112] Ibid.

CBD[113] and the SDGs,[114] to ensure coherence and synergy between trade policies and biodiversity objectives; and stakeholder engagement, including with Indigenous peoples with their wealth of traditional knowledge and practices related to the protection of biodiversity and biocultural heritage that have been accumulated over thousands of years. The effectiveness of FTAs in protecting biodiversity loss depends on the specific provisions' wording and their implementation and enforcement. To this end, it is essential to ensure regular monitoring and reporting, and to strike a balance between trade and conservation objectives to make certain that biodiversity and biocultural heritage are not compromised in the pursuit of economic gains. Some FTAs, such as the *US-Peru Free Trade Agreement*, allow for the protection of traditional knowledge and genetic resources.[115] Another example of an FTA that has provisions for protecting biodiversity, and has the potential to indirectly protect biocultural heritage as well, is the *CPTPP*, which includes provisions on traditional knowledge, genetic resources and biological conservation (as discussed in part five below).[116]

However, it is important to keep in mind that there are certain challenges attached to the protection of biodiversity and biocultural heritage by utilising FTAs. First and foremost, the extent to which FTAs can be used to protect biocultural heritage is limited by their focus on trade and economic issues, which may curb their effectiveness vis-à-vis ability to address broader environmental and cultural concerns related to biodiversity and biocultural heritage.[117] As has been observed, even where FTAs do include provisions for protecting biodiversity and biocultural heritage, their implementation and enforcement can be challenging.[118]

113 UN Environment Programme, 'The Global Biodiversity Framework – What's Next for Financial Policy and Regulation?' (Nature News, 1 February 2023). https://www.unepfi.org.

114 UN SDGs, 'A New Global Framework for Managing Nature Through 2030: 1st Draft Agreement Debuts'. https://www.un.org.

115 IIED, Biocultural Heritage, 'National and Local Policy and Law Protecting Biocultural Heritage' https://www.iied.org.

116 Harsh Bhati and Yaffa Epstein, 'Protection of Biocultural Heritage in the Anthropocene: Towards Reconciling Natural, Cultural, Tangible and Intangible Heritage' (2023) 20 *Journal of Environmental Law* 1.

117 Ibid.

118 IIED (Footnote 115).

Also, biocultural heritage is not always recognised or valued in the same way as other forms of heritage, which can constrain its protection under FTAs.[119] Notwithstanding this, the literature reveals a mounting pressure for FTAs to address issues related to climate change as well as other issues that are not strictly trade or investment-related concerns, including biodiversity.[120]

Trade and climate change-related questions are indivisible.[121] Even the ultimate international legal instrument concerned with international trade—the *Agreement establishing the World Trade Organization* ('*WTO Agreement*')—in its Preamble, explicitly recognises that its State Parties' relations in the field of trade and economic endeavour should be conducted in a manner that nations' desires for development and improving standard of living, including through international trade, should allow 'for the optimal use of the world's resources in accordance with the objective of sustainable development, seeking both to protect and preserve *the environment*'.[122] This is a paramount recognition of the inseparability of international trade and the environment, including the impacts of climate change. The preamble to a treaty sets the tone and spirit of a treaty and, consistent with Article 31(2) of the *Vienna Convention of the Law of Treaties*,[123] is considered an integral part of the treaty, together with the main text and annexes. While beyond the scope of this chapter to expand on this point further, it is suggested that the World Trade Organisation (WTO) should play a more prominent role in incorporating environment-related issues into its agenda and in ensuring that they are not overlooked at the FTAs level either.[124] There are many

[119] Ibid.

[120] Lee Rarrick, 'Biodiversity Impacts of Investment and Free Trade Agreements' (2019) 37(1) *Pace Environmental Law Review* 67; Dent (Footnote 16); Benson and Judd (Footnote 18).

[121] Jadranka Petrovic, 'Free Trade Agreements and International Trade' in Gerry Nagtzaam et al. (eds), *Legal Pathways to Deep Decarbonisation in Australia* (LexisNexis, 2023) 67.

[122] *Agreement establishing the World Trade Organisation*, 1867 UNTS 154, opened for signature 15 April 1994 (entered into force 1 January 1995) preamble para 1 [*WTO Agreement*] (emphasis added).

[123] *Vienna Convention on the Law of Treaties*, opened for signature 23 May 1969, 1155 UNTS 331 (entered into force 27 January 1980), art 31(2) [*Vienna Convention*].

[124] Petrovic (Footnote 121).

benefits to addressing the environment-related issues in WTO multilateral and plurilateral agreements, as well as in FTAs. They include the enforceability of obligations undertaken under such agreements, as opposed to commitments under climate law international instruments which seem to be largely discretionary and voluntary in nature.

5.5 The Comprehensive and Progressive Agreement for Transpacific Partnership, and Biodiversity and Biocultural Heritage—the Australian Example

5.5.1 Australia's Legal Obligations vis-à-vis the Environment

Australia is particularly vulnerable to the consequences of climate change.[125] Climate change is especially devastating Indigenous Peoples' lands and waters, threatened species, heritage sites and other matters of environmental significance.[126] It has been argued that climate change impacts on the environment will be worsened by the coal and gas projects and the Australian Government has been asked to reconsider projects approvals under federal environmental law and to consider the broader effects of climate change and how emissions from such projects could damage the environment.[127]

Australia is bound by many international legal instruments which provide, to varying degrees, protection for biodiversity and heritage-related matters of concern to the international community. Key among these instruments include the 1994 *General Agreement on Tariffs and Trade* ('*GATT*'),[128] annexed to the *WTO Agreement*, whose Article 20 aims to protect human, animal or plant life or health, as well as non-renewable natural resources; the *CBD*—a specialist treaty—concerned with the international protection of biological diversity exclusively; the

[125] Mark Ludlow, 'Australia's Extreme Climate Leaves It Vulnerable to Global Warming' (*Financial Review*, 10 August 2021). https://www.afr.com.

[126] Sarah Sedghi, 'Labor Asked to Reassess Decisions about Coal and Gas Projects Made by Previous Governments' (BBC News, 11 July 2022). https://www.bbc.com/news.

[127] Ibid.

[128] *General Agreement on Tariffs and Trade*, 55 UNTS 194, opened for signature 14 April 1994 (entered into force 1 January 1995) art 20 [GATT].

CITES, whose goal is to protect endangered species; the *Convention on the Conservation of Migratory Species of Wild Animals* ('*CMS*'), which provides protection to migratory animals and habitats; the *Ramsar Convention on Wetlands*, aimed at supporting healthy wetland ecosystems and the *World Heritage Convention*, which focusses on preserving natural sites, cultural properties and mixed properties. Implementation of obligations emanating from international instruments occurs at the domestic level.

At the domestic level, the central piece of legislation on the subject of the protection of the environment and biodiversity, as well as heritage, is the *Environment Protection and Biodiversity Conservation Act 1999* (Cth) ('*EPBC Act*').[129] The *EPBC Act* covers a wide spectrum of matters of national environmental significance, including nationally threatened species and ecological communities; migratory species; World Heritage properties; National Heritage places; wetlands of international importance (often referred to as 'Ramsar' wetlands after the treaty under which such wetlands are listed); Commonwealth marine areas, the Great Barrier Reef Marine Park; nuclear actions (including uranium mining) and a water resource, in relation to coal seam gas development and large coal mining development.

The *EPBC Act* provides protection to Indigenous heritage through National Heritage or World Heritage listing. At the national level, Indigenous cultural heritage is protected under numerous other Commonwealth laws, including the *Aboriginal and Torres Strait Islander Heritage Protection Act 1984* ('*ATSIHP Act*').[130]

In the past several years the *EPBC Act* has undergone considerable scrutiny. The statutory review of this act commenced on 29 October 2019, with the appointment of Professor Graeme Samuel AC as the independent reviewer. On 30 October 2020, Professor Samuel provided the Minister for the Environment with his *Final Report*. The *Report* was tabled in Parliament on 28 January 2021.[131] The *Report* is comprehensive, containing 12 chapters and appendices. The *Report* found that, *inter alia*: the *EPBC Act* is outdated and that it does not effectively address

[129] *Environment Protection and Biodiversity Conservation Act 1999* (Cth) ['*EPBC Act*'].

[130] *Aboriginal and Torres Strait Islander Heritage Protection Act 1984* ['*ATSIHP Act*'].

[131] *Independent Review of the Environment Protection and Biodiversity Conservation Act 1999 (Cth): Final Report*, October 2020. https://www.environment.gov.au.

key threats to the environment and where addressing the challenge of adapting to climate change this is an implied, rather than a central consideration; although significant efforts are made to assess and list threatened species, once listed, not enough is done to deliver improved outcomes for them; Western science is heavily prioritised in the way the *EPBC Act* operates whereas Indigenous knowledge and views are diluted in the formal provision of advice to decision-makers, which reflects an overall culture of tokenism and symbolism, rather than one of genuine inclusion of Indigenous Australians; and that fundamental reform of national environmental laws is required.[132] The *Report* recommended urgent adoption of a number of standards, including those related to Australia's First Nations peoples, characterising them as urgent and placing them at the centre.[133]

On 8 December 2022, the Federal Government released its response to *Professor Samuel's 2020 Review of the EPBC Act*, titled, *Nature Positive Plan: Better for the Environment, Better for Business*.[134] In the Government's response climate change does not figure highly. The Government is not proposing as part of this response to introduce climate change trigger into the *EPBC Act*. The *Plan* suggests that the Government does not intend to use its power to approve or refuse projects as a mechanism to meet its emission reduction targets. In relation to First Nations peoples' cultural heritage the Government will be developing, as a priority, a National Environmental Standard for First Nations Engagement and Participation in Decision-Making, however, the Government has provided little detail about what this standard will entail.[135] This commitment vis-à-vis First Nations Australians' engagement and participation in decision-making is a step in the right direction—a humble step towards improving the protection of the nature on the Australian continent, but steps forward, nevertheless. But how does Australia fare in this respect in the international trade context, specifically, under the CPTPP?

[132] Ibid.

[133] Ibid.

[134] DCCEEW, 'Nature Positive Plan: Better for the Environment, Better for Business' (2022). https://www.environment.gov.au.

[135] Bill McCredie et al., 'Government Announces Major Reforms to EPBC Act' (9 December 2022). https://www.allens.com.au/insights.

5.5.2 *Comprehensive and Progressive Agreement for Transpacific Partnership: Environment Chapter*

It is worth reiterating that the *CPTPP* was signed by 11 States, including Australia, on 8 March 2018, and entered into force on 30 December 2018.[136] The text of its predecessor legal instrument, the *Trans-Pacific Partnership* ('*TPP*')—which was signed in February 2016—with just a few slight amendments, became the text of the *CPTPP*.[137] The *CPTPP* dedicates an entire chapter to the environment. The *CPTPP* Environment Chapter (Chapter 20) is an enforceable[138] and comprehensive chapter that houses 23 articles. While neither the *CPTPP* generally, nor its Environment Chapter specifically, contain explicit provisions concerning biocultural heritage, Chapter 20 does include provisions promoting sustainable development,[139] trade and biodiversity,[140] and provisions concerning conservation and trade which emphasise the illegal trade in wild fauna and flora,[141] as well as provisions dealing with trade in environmental goods and services.[142] However, a closer examination of the text reveals that State Parties are subject to little external oversight when making decisions regarding environmental issues, and can remain presumptively discretionary in regard to the implementation of policies designed around specific applications of any given clause. This discretion can result in a diverse response to an issue of everyone's concern, which can lead to inconsistency and unpredictability regarding pressing environmental matters. In this sense, the issues of flora and fauna, and the question of biodiversity warrant attention in the context of Australia.

When examining integrated policies with relevance to the conservation of wild flora and fauna, particularly the illegal importing and exporting of endangered species, it should be noted that the wording of the *CPTPP*

[136] *CPTPP* (Footnote 28).

[137] The text of the *CPTPP* is available at Australia's DFAT website at. https://www.dfat.gov.au.

[138] *CPTPP* art 20.23 (dealing with dispute resolution relevant to Article 20 matters) and art 28 concerning an overall Dispute Settlement.

[139] *CPTPP* art 20.2.

[140] *CPTPP* art 20.13.

[141] *CPTPP* art 20.17.

[142] *CPTPP* art 20.18.

Environment Chapter remains weak and ultimately offers little protection. Article 20.17 of the chapter, which is concerned with 'conservation and trade' appears to be somewhat of a paper tiger. Paragraph (2) of this article imposes the 'obligation' on State Parties to fulfil existing obligations under the *CITES* (the above-mentioned leading international legal instrument concerning international trade in endangered species of wild fauna and flora).[143] The remainder of Article 20.17 fails to prohibit the illicit trade in a meaningful way. Instead, it only 'encourages' parties to combat the illegal trade in wild flora and fauna in a non-binding manner. Article 20.17(3) extends on this by requiring parties to 'promote' conservation as well as take action to fight the illicit trade.[144] While the trading market for flora and fauna within Australia is relatively small in comparison with that of other State Parties, a two-way trading system does exist with imports and exports causing harm to Australian fauna and flora.[145]

Language similarly weak and likely to be unenforceable is echoed in the provision of Article 20.13 covering trade and biodiversity where the State Parties 'recognise the importance of conservation and sustainable use of biological diversity and their key role in achieving sustainable development'.[146] This article goes on in a similar manner in its paragraph (2) committing the State Parties to 'promote' and to 'encourage' the conservation and sustainable use of biological diversity.[147] The operational language used in this paragraph, and elsewhere in general leaves the *CPTPP* Environment Chapter looking vague and lacking any compelling character. Australia has been included in a short list of States being described as 'megadiverse', holding one of the richest and most distinct assortments of flora and fauna in the world.[148] Today Australia is experiencing the greatest decline in biodiversity of any continent to ever be

[143] *CPTPP* art 20.17(2).

[144] *CPTPP* art 20.17(3).

[145] Carolyn Barry, Australia's Wildlife Black Market Trade' (2011) *Australian Geographic*. https://www.australiangeographic.com.au. In Australia, the maximum penalty for wildlife trade offences is 10 years imprisonment and/or fine of up to $210.000.

[146] *CPTPP* art 20.13(1).

[147] *CPTPP* art 20.13(2).

[148] Darcie Carruthers, 'Australia is a Megadiverse Country—What Does It Actually Mean?' (*Nature*, 7 February 2023). https://www.acf.org.au.

documented, with a large number of species under threat.[149] Given that Australia and some other *CPTPP* State Parties host some of the world's biodiversity hotspots, this FTA would benefit from improved and more meaningfully enforced environmental protection.

Moreover, and worryingly, despite the alarming impact of climate change on the environment, including its devastating effect on biocultural diversity—the most serious threat that the planet Earth is facing currently—there is not a single reference to climate change anywhere in this agreement. Nor is there unequivocal reference to the international legal instruments concerned with climate change, namely, the *UNFCCC*[150] and the *Paris Agreement*[151]—to both of which all parties to the *CPTPP* are party as well. The *CPTPP* has been subject to heavy criticism because of this omission[152] and because it falls woefully short in terms of climate action (especially because the draft *TPP* did unambiguously mention the climate change phenomenon, but the explicit reference to climate change was later removed and did not appear in the 2016 signed version of the *TPP* which, as noted, became integral part of the 2018 *CPTPP*).[153] How does the CPTPP respond to the protection of the environment at the practical level in the Australian milieu?

5.5.3 The Great Barrier Reef

The Australian continent is highly vulnerable to the consequences of climate change. Its impacts are especially being observed throughout the Great Barrier Reef, and are in fact the biggest threat to its survival.[154] The Great Barrier Reef is one of the richest and most complex natural

[149] Tracey Ferrier, 'Five Australian Species at Immediate Risk of Extinction' (*The Canberra Times*, 17 August 2023). https://www.canberratimes.com.au.

[150] *UNFCCC* (n10).

[151] *Paris Agreement* (Footnote 72).

[152] Matthew Rimmer, 'Greenwashing the Trans-Pacific Partnership: Fossil Fuels, the Environment, and Climate Change' (2016) 14 *Santa Clara Journal of International Law* 488.

[153] Karen Barlow, 'Robb on the TPP: "Read the Fine Print"' (The Huffington Post, 5 November 2015). https://www.huffingtonpost.com.au/2015/11/05/robb-on-the-tpp-read-the-fine-print.

[154] Great Barrier Reef Foundation, 'Climate Change is the Single Biggest Threat Facing the Reef'. https://www.barrierreef.org-reef/threats/climate-change.

ecosystems on the planet, but climate change is the number one danger, manifested in the form of coral bleaching, ocean acidification, severe weather events and habitat changes.[155]

Research shows that the Great Barrier Reef is an important part of local Indigenous Peoples' history, culture and spirituality.[156] Australia's First Nations peoples have known and explored the Reef for tens of thousands of years.[157] The Great Barrier Reef has been an integral part of the cultural heritage of these communities. The cultural significance of the Great Barrier Reef to First Nations peoples has been largely documented through their paintings, songs and stories that have been passed down through generations.[158] The Reef also provides Indigenous Australians with sustenance and resources for their livelihoods.[159] In addition to being a source of food, the Reef has also served as a gathering place and a spiritual site.[160] Research further reveals that First Nations peoples have a deep understanding of the Reef's ecology and that they have developed sustainable practices for managing its resources.[161] The contemporary management and conservation of the Great Barrier Reef

155 Ibid.

156 Australian Government, Great Barrier Reef Marine Park Authority [GBRMPA], *Strong Peoples—Strong Country Indigenous Heritage Monitoring Framework—Summary Report*, 2019. https://www.gbrmpa.gov.au; Australian Government, *Traditional Owners of the Great Barrier Reef: The Next Generation of Reef 2050 Actions—Final Report from The Reef 2050 Traditional Owners Aspirations Project*, December 2018. http://www.agriculture.gov.au.

157 Josephine Flood, *The Original Australians: The Story of the Aboriginal People* (Allen & Unwin, rev and updated ed, 2019); GBRMPA, 'Traditional Owners of the Great Barrier Reef' (Reef Facts Issue 4). http://www.gbrmpa.gov.au.

158 Ilisapeci Lyons et al., 'Putting Uncertainty under the Cultural Lens of Traditional Owners from the Great Barrier Reef Catchments' (2019) 19 *Regional Environmental Change* 1597.

159 Australian Government, *Reef 2050 Long-Term Sustainability Plan 2021–2025* (2021). https://www.dcceew.gov.au; James Ford et al., 'The Resilience of Indigenous Peoples to Environmental Change' (2020) 2(6) *One Earth* 532.

160 Christina Hicks et al., 'Securing Local Aquatic Food Systems in the Face of Declining Coral Reefs' (2021) 4(9) *One Earth* 1214.

161 Tony Jupp and Renee Woods, 'How Humans Have Impacted Australia's Nature over Time: A Brief History of the Impacts of Indigenous and Non-Indigenous Australians on Nature' (The Nature Conservancy Australia, 2 October 2019). https://www.natureaustralia.org.au.

has been increasingly recognising Indigenous Peoples' knowledge and practices.[162]

The Reef also has important economic value in the tourism domain as it is a popular tourist destination and as such is an important monetary resource for Indigenous Peoples and other local communities.[163] It contributes more than AU$6.4 billion each year to the Australian economy and supports about 64,000 full-time jobs.[164]

The importance of the protection of the Great Barrier Reef transcends local and national boundaries because the Reef is a World Heritage site inscribed on the UNESCO's World Heritage List in 1981 because of its outstanding universal value including its unique natural attributes and enormous scientific and environmental significance.[165] As such its preservation is of the importance to all humankind.[166]

The Reef has recently suffered severe coral bleaching (there had been four mass coral bleaching events since 2016 until this writing)[167] due to an increase in water temperature, which has affected an immense variety of organisms that rely on and live off these corals.[168] Additionally, and interrelatedly, resource extractive activities taking place in the vicinity of the Great Barrier Reef have been amplifying the impacts of climate change on the Reef.[169] Coal mining (which causes major carbon

[162] GBRMPA, 'Strategies to Manage the Reef'. https://www.gbrmpa; DCCEEW, 'Funding Boost for Traditional Owner-Led Reef Projects' (25 September 2023). https://www.dcceew.gov.au.

[163] Great Barrier Reef Organization, 'Record Investment in the Great Barrier Reef to Drive Jobs' (Australian Government Media Release, 29 April 2018). https://www.barrierreef.org/news/record-investment-in-the-great....

[164] Australian Government, 'About the Great Barrier Reef'. https://dcceew.gov.au/parks-heritage/great-barrier-reef/...

[165] UNESCO, 'UNESCO World Heritage Sites: Great Barrier Reef'. https://whc.unesco.org/en/list/154.

[166] *World Heritage Convention*, pream paras 6, 7.

[167] Great Barrier Reef Foundation, 'Mass Bleaching Events'. https://www.barrierreef.org/the-reef/threats/coral-bleaching.

[168] Hamish McGowan and Alison Theobald, 'Atypical Weather Patterns Cause Coral Bleaching on the Great Barrier Reef, Australia during the 2021–2022 La Nina' (2023) 13 *Scientific Reports* 6397.

[169] A. Saint-Amand et al., 'Quantifying the Environmental Impact of a Major Coal Mine Project on the Adjacent Great Barrier Reef Ecosystems' (2022) 179 *Marine Pollution Bulletin* 11365.

dioxide (CO_2) emissions and is thus linked to climate change and global warming both domestically and internationally through its export to other countries)[170] and hydraulic fracturing—the process of extracting gas by injecting fluids including water, sand and chemicals under pressure into coal seams to force open existing fissures—are impacting negatively on flora and fauna.[171] Recently, challenges surrounding the Great Barrier Reef almost resulted in degrading its status from the World Heritage List to the World Heritage in Danger List.[172] The Australian Government has been making efforts to step up the protection of the Reef. Recently, for the first time in history, Australia has rejected a coal mine project under its federal environmental laws as posing an unacceptable risk to the World Heritage area, which is already highly vulnerable.[173] However, despite this specific positive development, research shows that Australia—the third largest exporter of fossil fuels in the world—is still approving new coal mines or expansions. According to the Australian Institute's Coal Mine Tracker, four such projects have been approved by the Federal Environment Minister since May 2022 with 147 million tonnes of lifetime emissions,[174] which goes against its climate goals. Research further reveals that there are still 19 coal and gas projects which the Federal Environment Minister has been asked to reconsider vis-à-vis the broader effects of climate change and the question of how emissions from these projects could affect the environment.[175] All in all, ignoring climate change—which is caused, *inter alia*, by the harmful coal and other fossil fuel-related practices—in mega agreements such as the *CPTPP* can only exacerbate damage to the already vulnerable Great Barrier Reef.

[170] Frances Mao, 'Climate Change: Why Australia Refuses to Give up Coal' (BBC News, 22 October 2021). https://www.bbc.com/news.

[171] Friends of the Earth, *Fracking the Planet: How the Trans-Pacific Partnership Will Expand Fracking in Australia and around the Globe* (Friends of the Earth, 2015). https://www.foe.org.au.

[172] Christopher Testa and Conor Byrne, 'World Heritage Committee Keeps Great Barrier Reef off UNESCO List of Sites in Danger' (ABC News, 14 September 2023). https://www.abc.net.au; UNESCO *World Heritage Convention*, 'Great Barrier Reef'. https://whc.unesco.org/en/list/154.

[173] Tiffanie Turnbull, 'Australia Blocks Coal Mine to Protect Great Barrier Reef' (BBC News, 8 February 2023). https://www.bbc.com/news.

[174] Australian Institute, 'Coal Mine Tracker'. https://australianinstitute.org/au.

[175] Sedghi (Footnote 126).

5.5.4 *Indigenous Peoples and Biocultural Heritage*

The *CPTPP* does not make direct reference to Indigenous Peoples in relation to biocultural diversity and biocultural heritage. However, it does address the importance of cooperation in the area of Indigenous Peoples' traditional knowledge in Article 16 of its Chapter 18 concerned with Intellectual Property.[176] The *CPTPP* also reaffirms in its Preamble the importance of promoting Indigenous rights, sustainable development and traditional knowledge, and of preserving their right to regulate in the public interest.[177] Laudably, the CPTPP is the first regional trade agreement to recognise Indigenous Peoples' rights in its preamble.[178] The Preamble of the *CPTPP* sets the tone and spirit for the implementation of this treaty and provides space for the pursuit of more equitable, fair and inclusive trade.[179] However, as Harrison and Asgari correctly observe, the degree of inclusion and recognition of Indigenous Peoples and their rights differs significantly across *CPTPP* Member States.[180]

It has been rightly noted by some that recognising and upholding Indigenous Peoples' rights in international trade requires an examination of the intersections between multilateral trade and Indigenous Peoples' lands, resources, knowledge and cultural heritage to provide for inclusion, equity and protection of Indigenous Peoples' rights.[181] This is undoubtedly a significant area that not only warrants the attention of governments and various other players at the domestic level, but also demands the WTO's consideration.

[176] *CPTPP* art 18.16.

[177] *CPTPP* pream para 6.

[178] Scott Harrison and Robin Asgari, 'International Trade Inclusivity: The *CPTPP* and Indigenous International Trade and Investment' Asia Pacific Foundation of Canada (17 December 2020). https://www.asiapacific.ca; Risa Schwartz and Judy Whiteduck, 'A Proposal for a Joint Declaration on Trade and Indigenous Peoples' (Centre for International Governance and Innovation, 27 April 2020). https://www.cigionline.org.

[179] Harrison and Asgari (Footnote 178).

[180] Ibid.

[181] Schwartz and Whiteduck (Footnote 178).

5.6 Way Forward and Concluding Remarks

Climate change and biodiversity loss are interwoven, complex and multifaceted crises that necessitate urgent attention. A multiscalar problem requires a multiscale approaches and strategies in all domains. One way to contribute to ensuring a just transition to net-zero is by utilising FTAs and the involvement of those affected by the transition, including marginalised and vulnerable Indigenous Peoples whose millennia-long traditional knowledge and practices and deep connections with nature should be deemed invaluable in the protection of biodiversity and biocultural heritage. In its recent decision related to seismic ocean tests the Federal Court of Australia confirmed the importance of consulting Indigenous Peoples on these matters.[182] There is limited information specifically on the impact of a just transition on biocultural diversity and there is the need for further research vis-à-vis the impacts of FTAs on biocultural heritage sites. There are also potential challenges associated with the relationship between FTAs and the protection of the natural environment and biocultural diversity, including FTAs' focus on trade and economic issues which may limit their ability to address broader environmental and cultural concerns. Notwithstanding these issues there is growing recognition of the importance of biocultural heritage for sustainable development, and efforts are underway to promote its protection through national policies, and laws both national and international, as well as an ever-increasing call for FTAs in particular to contribute in this regard.

It has been argued in this chapter that despite potential impediments, FTAs do have the capacity to contribute to achieving the *Paris Agreement*'s 1.5 °C target. To this end, the involvement of Indigenous Peoples in the negotiation processes and various stages of a just transition to decarbonisation is a *conditio sine qua non*. Moreover, to ensure meaningful protection of the natural environment and biocultural heritage, FTAs need to be designed in a manner that confirms the importance of integrating the just transition to decarbonisation, highlighting first and foremost the urgency of phasing out coal mining—the main culprit in the context of climate change and its resultant harmful effects on the environment. To achieve this aim, FTAs need to incorporate carefully drafted,

[182] Angela Macdonald-Smith, 'Indigenous Win on Woodside's Scarborough Testing Approvals' (*Financial Review*, 28 September 2023). https://www.afr.com.

unambiguous and enforceable provisions. Importantly, plans should be put in place to guarantee regular reporting and monitoring compliance at the practical level. All stakeholders should be involved in the relevant processes, where Indigenous Peoples and civil society could and should play a vital role. This is another area worthy of future research. Further, a synergistic approach to the climate crisis and the interrelated biocultural crisis is required across the relevant international legal instruments to achieve harmonious effectiveness. Equally imperative is that the WTO becomes more active in reminding its Membership of the necessity of translating this organisation's aims and objectives related to the indivisibility of international trade and sustainability and the protection of environment, which have been impressively pledged for in its constitutive legal instrument. Hopefully, this and the associated trade-environment-related key issues will be taken into consideration during the COP28 climate discussions. Finally, each inhabitant of the plant Earth can, and should, contribute to the conservation of biodiversity and the protection of biocultural heritage in many ways, including reducing their own carbon footprints by adopting sustainable practices in their daily lives; and by educating themselves and others about the importance of preserving biodiversity and biocultural heritage and the threats they face,[183] and about the paramountcy of transitioning to net-zero in a way that is not only sustainable but also just and equitable so that our succeeding generations will not be deprived of what we have inherited.

[183] UNESCO, 'Climate Change Education'. https://www.unesco.org/en/climate-change/education.

CHAPTER 6

Youth Empowerment for a Just Transition

Abstract Youth are potentially powerful advocates for climate justice and the key to unlocking a just transition to a low-carbon economy; however, they must be empowered as the custodians of the future. Article 12 of the *Paris Agreement* recognises that education, training and public participation are foundational to achieving climate action. This is because youth and members of the public must be appropriately equipped with the knowledge, skills and literacy to create equitable system-wide transformation across all sectors. This is important to hold governments to account and determine the future, as these individuals take their place as the next generation of leaders, managers, decision-makers and voters. Education and youth participation can drive a just transition, and we discuss the barriers and enablers. This chapter explores the history of the *Action for Climate Empowerment (ACE)* agenda. It explains the six elements of Article 6 of the *United Nations Framework Convention on Climate Change*, which are education, training, public awareness, public participation, public access to information and international cooperation.

Keywords Just transition · Youth education · Voice of youth · Action for Climate Empowerment · Paris Agreement · Climate justice

D. Kraal et al., *A Just Transition to Decarbonisation*, Just Transitions,
https://doi.org/10.1007/978-3-031-49137-5_6

6.1 Introduction

Over the past 10 years the cascade of climate change impacts anticipated by the Intergovernmental Panel on Climate Change (IPCC)[1] has become a stark reality. Acute extreme climatic events like wildfires, droughts, floods and heatwaves are increasingly common around the globe along with more gradual impacts such as rising sea levels. These consequences have severe outcomes including injury, illness and loss of life, with disproportionate impacts on the world's most vulnerable.[2,3]

The 2023 IPCC synthesis report[4] emphasises the urgent need for more ambitious national and global action. Unfortunately, national governments are taking inadequate decarbonisation measures.[5] As things stand, our current policies make it increasingly likely that global warming will exceed 1.5°C within this century, making it significantly more challenging to restrict the temperature rise to below 2°C as agreed in the *Paris Agreement*. Executive Secretary of United Nations Climate Change Simon Stiell, asserts:

> We are still nowhere near the scale and pace of emission reductions required to put us on track toward a 1.5 degrees Celsius world. To keep this goal alive, national governments need to strengthen their climate action plans now and implement them in the next eight years.[6]

[1] Intergovernmental Panel on Climate Change (IPCC). https://www.ipcc.ch/.

[2] Perry E Sheffield and Phillip J Landrigan, 'Global Climate Change and Children's Health: Threats and Strategies for Prevention' (2011) 119 (3) *Environmental Health Perspectives* 291.

[3] World Health Organization (WHO), 'WHO Issues Urgent Call for Global Climate Action to Create Resilient and Sustainable Health Systems'. https://www.who.int/news/item/24-05-2023-wha76-strategic-roundtable-on-health-and-climate. Accessed 2023.

[4] IPCC, 'Summary for Policy Makers' in H Lee and J Romero (Core Writing Team, eds.) *Climate Change 2023: Synthesis Report. Contribution of Working Groups I, II and III to the Sixth Assessment Report of the Intergovernmental Plan on Climate Change* (IPCC, 2023).

[5] Ibid.

[6] United Nations Climate Change (UN Climate Change), 'Climate Plans Remain Insufficient: More Ambitious Action Needed Now' (Press Release 26 October 2022). https://unfccc.int/news/climate-plans-remain-insufficient-more-ambitious-action-needed-now.

The *Emissions Gap Report 2002: The Closing Window*[7] stresses the necessity of urgent system-wide transformation. Inger Anderson (Executive Director of the United Nations Environment Programme) argues that, 'We had our chance to make incremental changes, but that time is over. Only a root-and-branch transformation of our economies and societies can save us from accelerating climate disaster'.[8] This is likely to be made clear at the 28th Conference of the Parties (COP28) hosted in Dubai in 2023 during the first-ever global stocktake[9] on the *Paris Agreement*, which will create an inventory of the progress of countries and stakeholders to date.

The younger generation acknowledges that they will inherit a world burdened with even more profound climate impacts than that experienced by previous generations. Indeed, youth and other vulnerable demographics will disproportionately bear the brunt of the 'era of global boiling'.[10] For example, the World Bank has projected that climate change and the COVID pandemic may force an additional 150 million people into extreme poverty.[11] Further, estimates from the International Food Policy Research Institute suggest a 20% increase in malnourished children by 2050, compared with a scenario without climate change.[12] The concerning reality of the weight of the climate crisis is resulting in young individuals grappling with higher levels of climate anxiety and helplessness compared with the general population. A recent study highlights

[7] United Nations Environment Programme (UNEP), 'Emissions Gap Report 2022: The Closing Window: Climate Crisis Calls for Rapid Transformation of Societies' (UNEP, 2022). https://www.unep.org/resources/emissions-gap-report-2022.

[8] UNEP, 'Inadequate Progress on Climate Action Makes Rapid Transformation of Societies Only Option' (Press Release, 27 October 2022). https://www.unep.org/news-and-stories/press-release/inadequate-progress-climate-action-makes-rapid-transformation.

[9] UN Climate Change, 'Why the Global Stocktake is a Critical Moment for Climate Action'. https://unfccc.int/topics/global-stocktake/about-the-global-stocktake/why-the-global-stocktake-is-a-critical-moment-for-climate-action.

[10] 'Hottest July Ever Signals "Era of Global Boiling Has Arrived" Says UN Chief', *UN News*, July 2023. https://news.un.org/en/story/2023/07/1139162.

[11] The World Bank, 'COVID-19 to Add as Many as 150 Million Extreme Poor by 2021' (Press Release, 7 October 2020). https://www.worldbank.org/en/news/press-release/2020/10/07/covid-19-to-add-as-many-as-150-million-extreme-poor-by-2021.

[12] Gerald C Nelson et al., *Food Policy Report: Climate Change: Impact on Agriculture and Costs of Adaptation* (International Food Policy Research Institute, 2009).

that a staggering 75% of individuals aged 16–25 years admit to feeling fearful about the future because of climate change.[13,14]

Children and youth will be impacted by environmental changes and environmental determinants of health in ways current and previous generations have not experienced. In 2020, the Human Rights Council drew attention to the need to realise the rights of younger generations to a healthy environment in both international and national policies. The World Health Organization frames this as a human rights issue as well as an issue of intergenerational equity, stating that the rights, development and wellbeing of youth should be better reflected in state implementation of the *United Nations Framework Convention on Climate Change* ('*UNFCCC*') in accordance with the Preamble of the *Paris Agreement*.[15,16,17] Therefore, this section highlights that young peoples' futures are interlinked with climate justice and the quest for a just transition to decarbonisation.

6.1.1 *Youth as Champions of Climate Justice*

Within this context, the climate crisis has prompted a notable surge in the influence of young people who are fervently advocating for urgent global action. A diversity of activists and rights-based litigators (at least one quarter of all plaintiffs in climate litigation are young people)[18] are pushing governments to commit to decarbonisation to protect Earth's

13 Janis Whitlock, 'Climate Change Anxiety in Young People' (2023) 1 *Nature Mental Health* 297. https://doi.org/10.1038/s44220-023-00059-3.

14 Laelia Benoit, Isaiah Thomas and Andrew Martin, 'Review: Ecological Awareness, Anxiety, and Actions Among Youth and Their Partners—a Qualitative Study of Newspaper Narratives' (2022) 27(1) *Child and Adolescent Mental Health*. https://doi.org/:10.1111/cambh.12514.

15 WHO, *Policy Brief: Climate Change, Health and Intergenerational Equity* (WHO Climate Change and Health Unit, 2022). https://www.who.int/publications/m/item/who-policy-brief--climate-change--health---intergenerational-equity.

16 UN Climate Change, 'The Paris Agreement'. https://unfccc.int/process-and-meetings/the-paris-agreement.

17 UN Climate Change, 'The Paris Agreement Preamble'. https://unfccc.int/sites/default/files/english_paris_agreement.pdf.

18 Elisabeth Donger, 'Children and Youth in Strategic Climate Litigation: Advancing Rights through Legal Argument and Legal Mobilization' (2022) 11(2) *Transnational Environmental Law* 263. https://doi.org/:10.1017/S2047102522000218.

essential life support systems and ensure human wellbeing. Despite lacking formal political and decision-making authority, youth are leveraging public demonstrations as a means of exerting pressure. Worldwide, millions of young people are expressing their concerns and demanding governmental action in response to the escalating climate emergency, including through harnessing social media.[19]

As a feature of this global movement, youth are resolutely championing the cause of climate justice. The United Nations Children's Fund (UNICEF) defines climate justice as an integration of human rights and development, adopting a people-centric approach to address the climate crisis.[20] It recognises the disparate global contributions to climate change and tackles various forms of inequality, encompassing social, gender, economic, intergenerational and environmental imbalances. As our physical, social and economic landscapes shift from historical norms, the most vulnerable groups—including youth, women and girls, persons with disabilities, Indigenous communities, the LGBTIQ+ population and those in impoverished and conflict-affected regions—will bear the brunt of impacts. This is because these groups are at a higher risk of discrimination, poverty and violence, and may not have access to resources and the capacity to adapt to changing social and environmental conditions.

Youth are taking unprecedented steps to confront these specific challenges, possessing a keen awareness of the intrinsic connection between global climate action and global inequality, particularly concerning the predicaments faced by developing nations.[21] Youth including Indigenous youth from nations on the frontline of the climate crises, such as Small Island Developing States, have been particularly influential in amplifying their unique cultural and regional perspectives and rights to

[19] Shelly Boulianne, Mireille Lalancette and David Ilkiw, 'School Strike 4 Climate: Social Media and the International Youth Protest on Climate Change' (2020) 8(2) *Media and Communication* 208. https://doi.org/:10.17645/mac.v8i2.2768.

[20] UNICEF, *Recap and Reflections: Climate Justice Roundtable* (Office of Global Insight and Policy, 2022). https://www.unicef.org/globalinsight/media/2866/file.

[21] United Nations Office of the High Representative for the Least Developed Countries, Landlocked Developing Countries and Small Island Developing States, 'On the Frontline of Climate Crisis, Worlds Most Vulnerable Nations Suffer Disproportionately'. https://www.un.org/ohrlls/news/frontline-climate-crisis-worlds-most-vulnerable-nations-suffer-disproportionately.

self-determination.[22] Despite minimal national contribution to emissions, many developing nations bear a disproportionate burden, compelling young leaders to call on developed countries for more substantial support in mitigating the increasingly evident consequences of extreme weather events and rising sea levels, and the existential threats these pose to communities and cultures. Eric Njuguna, a youth climate justice organiser, stated:

> It is widely acknowledged that those who played a minor role in causing the climate crisis are the ones suffering its harshest consequences, and regrettably, climate justice is a topic that often remains neglected.[23]

Youth have been vocal advocates on this matter at the annual Conference of the Parties (COP) of the *UNFCCC*, the preeminent global forum for climate action.[24] For example, the official youth constituency of the *UNFCCC* has a specific human rights workshop group. Further, the 2022 *COP27 Global Youth Statement* ('*Global Youth Statement*') from the 17th UN Climate Change Conference of the Youth (COY)[25] is subtitled *Declaration for Climate Justice*. It was formed via consultation with young people from 149 countries and calls 'for an inclusive approach to climate governance', finance to 'enable the most affected people and areas and climate-vulnerable countries' and 'justice in the transformations ahead'.[26]

Related to this, youth have focussed on the issue of loss and damage[27] at recent COPs. This issue acknowledges that affluent nations should

[22] United Nations Department of Economic and Social Affairs, 'Indigenous Youth as Agents of Change for Self-Determination 2023'. https://social.desa.un.org/issues/indigenous-peoples/events/international-day-of-the-worlds-indigenous-peoples-2023.

[23] UNICEF, *Recap and Reflections* (Footnote 18).

[24] UN Climate Change, 'Everything You Wanted to Know about COP but Were Too Afraid to Ask' (Blog Post, 18 October 2021). https://unfccc.int/blog/everything-you-wanted-to-know-about-cop-but-were-afraid-to-ask.

[25] COY17, 'The 17th UN Climate Change Conference of Youth COY17, Egypt 2022.' https://coy17eg.com/.

[26] COY17 and United Nations Institute for Training and Research; Children and Youth Constituency of the *UNFCCC* (YOUNGO), *COP27 Global Youth Statement: Declaration for Climate Justice* (YOUNGO, 2022) https://coy17eg.com/globalyouthstatement/.

[27] United Nations Climate Action, 'Transitional Committee on Loss and Damage Holds Second Meeting in Bonn 2023'. https://unfccc.int/news/transitional-committee-on-loss-and-damage-holds-second-meeting-in-bonn.

provide compensation to developing counterparts for the catastrophic repercussions of unsustainable development (see Chapter 2 for a more detailed explanation of this issue). Loss and damage as it pertains to climate finance has arguably been the single largest cross-cutting issue of recent negotiations, particularly at COP26 (Glasgow) and COP27 (Sharm el-Sheikh). At COP27, a breakthrough was made when a dedicated fund and new funding arrangements for vulnerable countries was announced.[28] However, there are concerns among youth about when and how this fund will be operationalised in a just way.

As we strive for a just transition to decarbonisation, it becomes more essential to tackle the various dimensions of climate justice. These include intergenerational justice and other justice aspects explored in this book. As seen at COP28 in Dubai, a new generation is passionately championing a just transition. They are advocating for a circular economy, workers' rights, and the development of skills to nurture a green workforce. Their rationale stems from the recognition that education and enhancing workforce capabilities will play pivotal roles in the future.

6.1.2 The Importance of Climate Empowerment

Action for Climate Empowerment ('*ACE*')[29] is an international framework that aims to empower youth (and the public) to drive climate action. *ACE* recognises that youth require access to quality climate change education and the opportunity and support to scale up their efforts and utilise their skills to grapple with complex issues of climate change. By enabling personal and collective agency, *ACE* aims to combat the destructive potential of climate change by cultivating international understanding among youth in recognition of the impacts on future generations. The issues of climate empowerment and justice are highly interlinked. This is because for empowerment to occur, efforts must recognise and support equity, inclusion and diversity as well as the specific local contexts and human rights of individuals, including the world's most vulnerable peoples.

[28] UN Climate Change News, 'COP27 Reaches Breakthrough Agreement on New Loss and Damage Fund for Vulnerable Countries' (Press Release, 20 November 2022). https://unfccc.int/news/cop27-reaches-breakthrough-agreement-on-new-loss-and-damage-fund-for-vulnerable-countries.

[29] UN Climate Change, 'Action for Climate Empowerment'. https://unfccc.int/ace.

6.1.3 *Chapter Overview*

This chapter discusses ACE with regard to youth, climate justice and a just transition. It provides an overview of the history and development of the *ACE* agenda. It covers the six elements of Article 6 of the *UNFCCC*,[30] highlights the importance of education and concludes by calling for stronger mechanisms and support to enable meaningful youth participation in decision-making.

6.2 An Introduction to *Action for Climate Empowerment*

In recognition of its importance to the fight against climate change, *ACE* was integrated into the original 1992 *UNFCCC*[31] and is designated its own dedicated article within the *Paris Agreement*.[32] *ACE* embodies a transversal approach within public policies, aiming to nurture and unite capacities that promote transformative actions against climate change, acknowledging the profound and urgent nature of this challenge.

ACE calls on governments to establish and execute initiatives related to education and public awareness, provide training to individuals with expertise in science, technology and management, facilitate access to information and foster public engagement in addressing climate change and its consequences. These multifaceted and broad *ACE* endeavours are guided by objectives for the effective implementation of climate adaptation and mitigation, ultimately aligning with the overarching goal of the *UNFCCC*.[33]

However, *ACE* is yet to be implemented in a systematic way across nations. As countries are called upon to embed *ACE* strategies within their nationally determined contributions (NDCs), according to national

[30] *United Nations Framework Convention on Climate Change* ('*UNFCCC*'), 'United Nations Framework Convention on Climate Change'. https://unfccc.int/files/essential_background/background_publications_htmlpdf/application/pdf/conveng.pdf.

[31] Ibid.

[32] UN Climate Change, 'The Paris Agreement' (Footnote 15).

[33] United Nations Educational, Scientific and Cultural Organization (UNESCO), *Action for Climate Empowerment: Guidelines for Accelerating Solutions through Education, Training and Public Awareness* (UNESCO, 2016). https://unesdoc.unesco.org/ark:/48223/pf0000246435.

circumstances,[34,35] time will tell whether *ACE* can be effectively funded, coordinated and monitored in a meaningful and just way to meet the goals of the *Paris Agreement*.

6.2.1 The Six Elements

ACE serves as the cornerstone for a future that is both resilient to climate change and equitable. It consists of six closely linked components: education about climate change, training, fostering public awareness, encouraging public engagement, ensuring public access to information and promoting international collaboration across these aspects (as depicted in Fig. 6.1).[36,37] Recognising the intricate nature of the climate crisis, the effective implementation of all six elements of *ACE* is acknowledged as essential to enable youth (and the public) to both comprehend and actively contribute to addressing the climate emergency.

ACE recognises that education equips individuals with an understanding of the causes, effects and solutions of climate change, empowering them to make informed decisions and take appropriate individual and collective actions.[38] International collaboration is an important cross-cutting component. It can facilitate the exchange of best practices and lessons learned. Multilateral cooperation enables parties and organisations to mutually support one another for the development and implementation of *ACE*.

Training builds national and global capacity for a positive future, equipping individuals with the technical expertise and advanced knowledge essential for facilitating an industry- and community-based shift

[34] UN Climate Change, 'Nationally Determined Contributions (NDCs)'. https://unfccc.int/process-and-meetings/the-paris-agreement/nationally-determined-contributions-ndcs.

[35] UNESCO, *Integrating Action for Climate Empowerment into Nationally Determined Contributions: A Short Guide for Countries* (UNESCO, 2020). https://unesdoc.unesco.org/ark:/48223/pf0000373762?posInSet=1&queryId=e23e02bb-27a2-4f38-a696-231658a9f315.

[36] *UNFCCC* (Footnote 28).

[37] UN Climate Change, 'Action for Climate Empowerment Hub'. https://unfccc.int/ace-hub.

[38] UNESCO, *Action for Climate Empowerment: Guidelines* (Footnote 31).

Fig. 6.1 The Six *ACE* Elements (*Source* Included with permission from the *UNFCCC)*

towards environmentally sustainable economies and climate-resilient societies. Effective public awareness campaigns and outreach are necessary to engage diverse stakeholders in this collective endeavour, particularly regarding implementing both national and global climate change policies. This is closely related to public access to information: both aim to equip people with the knowledge and tools for active and informed participation. *ACE* also emphasises the importance of public engagement, including involvement in decision-making processes. These elements provide the scope to meet the objectives of *ACE* (Fig. 6.2).[39,40]

6.2.2 History and Development

ACE (Article 6) has been an integral part of the *UNFCCC* since its inception, when the convention's text was adopted in 1992.[41] The importance of international collaboration in achieving the objectives of Article 6 was highlighted in the 1997 *Kyoto Protocol* (Article 10). The significance of *ACE* is also evident in other international frameworks

[39] UNESCO, *Integrating Action for Climate Empowerment* (Footnote 33).

[40] Francesco Bassetti, 'Action for Climate Empowerment (Foresight)'. https://www.climateforesight.eu/seeds/action-for-climate-empowerment/.

[41] UNESCO, *Action for Climate Empowerment: Guidelines* (Footnote 31).

Scope	Objectives	
Education	Change habits in the long-term	Foster a better understanding of, and ability, to address climate change and its effects
Training	Develop practical skills	
Public Awareness	Reach people of all ages and walks of life	Promote community engagement, creativity and knowledge in finding climate change solutions
Public Access to information	Make information freely available	
Public Participation	Involve all stakeholders in decision-making and implementation	Engage all stakeholders in debate and partnership to respond collectively to climate change
International Cooperation	Strengthen cooperation, joint efforts and knowledge exchange	

Fig. 6.2 ACE Scope and Objectives[42,43] (*Source* Included with permission from the *UNFCCC)*

including the *Bali Guidelines* (2010), *Aarhus Convention* (2011), *Global Action Programme for Education for Sustainable Development* (2014), Sustainable Development Goals (SDGs, 2015) and *Escazú Agreement* (2018).[44]

Within the *UNFCCC* process, Article 6 is executed via work programmes. In 2002, during the eighth COP, the *New Delhi Work Programme*[45] was adopted to address the specific circumstances of countries in relation to progressing their national initiatives towards Article 6. In 2007 at COP13 (Bali), this programming was amended and extended for five years.

The eight-year *Doha Work Programme*[46] on Article 6 was adopted in 2012 at COP18 (Doha). This programme encouraged parties to appoint a National Focal Point for Article 6 and provide support, including technical, financial and access to information and materials. Parties also began

42 Ibid.

43 UNESCO, *Action for Climate Empowerment: Guidelines* (Footnote 31).

44 Ibid.

45 *UNFCCC*, 'New Delhi Work Programme on Article 6 of the Convention'. https://unfccc.int/cop8/latest/14_cpl3_sbstal23add1.pdf.

46 *UNFCCC*, 'Doha Work Programme on Article 6 of the Convention'. https://unfccc.int/resource/docs/2012/sbi/eng/l47.pdf.

hosting annual international dialogues to showcase and enhance relevant efforts.[47] These dialogues have served as a platform for governments, representatives of *UNFCCC*-established bodies, experts, practitioners and stakeholders to exchange experiences, best practices and learnings. It was at the third annual dialogue in June 2015 that *ACE* was adopted as a user-friendly term for referring to Article 6.

At COP21 (Paris) in 2015, governments agreed to cooperate in taking measures, as appropriate to their circumstances, to enhance climate change-related education, training, public awareness, public participation and public access to information, recognising the importance of these steps to enhance actions under the *Paris Agreement*.[48]

The *ACE* guidelines for accelerating solutions[49] were developed as a joint initiative between the *UNFCCC* and the United Nations Educational, Scientific and Cultural Organization (UNESCO), and launched in 2016 at COP22 (Morocco). They are intended to facilitate the work of National Focal Points to implement *ACE* activities in their countries. The guidelines provide a flexible, phased approach to the strategic and systematic implementation of *ACE* activities at the national level, driven by each country's circumstances.

The *Doha Work Programme on ACE* concluded in 2020, and its successor, the *Glasgow Work Programme*[50] was established at COP26 (2021). It recognises that *ACE* 'plays a key role in promoting the changes in lifestyles, attitudes and behaviours needed to foster low-emission, climate resilient and sustainable development'.[51]

The action plan for the *Glasgow Work Programme* on *ACE*[37] ('*ACE Action Plan*') was adopted in 2022 at COP27 in Sharm el-Sheikh, to provide a clear roadmap for the next four years of *ACE* activities. The plan provides a framework that is currently guiding parties to implement *ACE* both globally and within their respective nations. It includes suggested metrics and actions all parties could employ to implement *ACE*, and

[47] UN Climate Change, 'Dialogues on Action for Climate Empowerment'. https://unfccc.int/ace-dialogues.

[48] UN Climate Change, 'The Paris Agreement' (Footnote 15).

[49] UNESCO, *Action for Climate Empowerment: Guidelines* (Footnote 31).

[50] *UNFCCC, Glasgow Work Programme on Action for Climate Empowerment Sessional Proceedings Glasgow Climate Change Conference October/November 2021* (UNFCCC, 2021). https://unfccc.int/documents/310896.

[51] Ibid.

emphasises the need to appropriately finance *ACE* activities. It contains four priority areas for action[52]:

1. policy coherence
2. coordinated action
3. tools and support
4. monitoring, evaluation and reporting (MER).

At the 2023 Bonn intersessional (SB 58), the *ACE* Focal Points Academy focussed on MER,[53] which aims to strengthen 'the implementation of all six ACE elements at all levels, according to Parties' specific priorities, needs and national circumstances'.[54] The in-session annual dialogue included training and knowledge sharing on MER, including how MER intersects with the three other priority areas, and involved diverse stakeholders from multiple nations with guidance from the United Nations Institute for Training and Research; Children and Youth Constituency of the *UNFCCC* (YOUNGO) and the *Monitoring and Evaluating Climate Communication and Education Project*.[55]

6.2.3 *Action for Climate Empowerment for a Just Transition*

When *ACE* was guided by the *Doha Work Programme*, several of its guiding principles related to justice (Section B (14)).[56] For example, there were principles that guide the programme to take a 'gender and intergenerational approach' and 'an interdisciplinary multi-sectoral, multi-stakeholder and participatory approach' to the work of *ACE*.[57] A gender approach means ensuring that climate actions are gender responsive and promote women's participation in decision-making. An intergenerational

52 Ibid.

53 UN Climate Change, 'Action for Climate Empowerment' (Footnote 27).

54 UN Climate Change, 'Action Plan under the Glasgow Work Programme on Action for Climate Empowerment 2023'. https://unfccc.int/sites/default/files/resource/cop27_auv_ACE.pdf.

55 Monitoring and Evaluating Climate Communication and Education, 'The MECCE Project'. https://mecce.ca/, https://unfccc.int/resource/docs/2012/sbi/eng/l47.pdf.

56 *UNFCCC*, *Glasgow Work Programme* (Footnote 47).

57 Ibid.

approach refers to engaging people of all ages in finding solutions for climate change, taking into special consideration the vulnerabilities of youth and the elderly, who have a reduced capacity to cope independently.

Work leading to the formation of the *Glasgow Work Programme* highlighted further intersectional issues of justice; however, not all these ideas were included in the official document agreed to at COP26. A 2021 report entitled *Justice and Action for Climate Empowerment* suggested an increased focus on climate justice, arguing that:

> all people, including those often disenfranchised or marginalized, need to be involved if we are to ensure environmental justice is critical to how justice can be conceptualized within climate empowerment itself.[58]

Human rights were added to the text at COP27 to ensure those most vulnerable to climate change are represented. The 2022 *ACE Action Plan* under the *Glasgow Work Programme* on *ACE* acknowledges that:

> climate change is a common concern of humankind, Parties should, when taking action to address climate change, respect, promote and consider their respective obligations on human rights, the right to health, the rights of indigenous peoples, local communities, migrants, children, persons with disabilities and people in vulnerable situations and the right to development, as well as gender equality, empowerment of women and intergenerational equity.[59]

Thus, *ACE* is becoming more aligned with issues such as a just transition and loss and damage through including a broader conceptualisation of climate justice in relation to human rights. Moving forward, as the *ACE Action Plan* is implemented, there is general recognition from parties of the need to respect, promote and consider their respective obligations on these issues of climate justice.

[58] Debra Morrison et al., *Justice and Action for Climate Empowerment (ACE): Guidance for Equitable Capacity-Building and Justice-Centered Implementation Drafted for the UNFCCC Action for Climate Empowerment Activity Series 2021* (ECOS, 2021). tinyurl.com/ACEJusticeBrief.

[59] UN Climate Change, 'Action Plan under the Glasgow Work Programme' (Footnote 50).

6.2.4 *State of Play*

One of the most important activities included in the COP27 *ACE Action Plan* is the encouragement for parties to create a national *ACE* strategy. A national strategy, developed according to national circumstances, is important for aligning the financing of *ACE* activities with national budgets and planning, mapping existing *ACE* activities for the identification of key gaps and identifying national targets and indicators to monitor, evaluate and report upon to ensure progress. Parties are currently at various stages of development with regard to creating and implementing a national *ACE* strategy with consideration of the aforementioned principles. Developing a national strategy can help to support implementation of *ACE* at the national level by outlining tools, activities and processes to promote action on the six *ACE* elements.[60,61] The strategy will also enable the generation of appropriate policy instruments to ensure policy coherence, cross-sectoral and inclusive collaboration and stakeholder engagement on *ACE* at the national and sub-national level. As countries define their NDCs and *ACE* strategies, it is important they are taking a locally contextualised approach, with research suggesting that *ACE* is more impactful when it engages with justice and is responsive to the region.[62]

While much progress has been made on *ACE*, there are many challenges to confront regarding the design and delivery of national strategies. First, it is difficult to define key national initiatives within the broad scope of *ACE*. Because all six *ACE* elements are important and distinct from one another, it is difficult for countries to identify the desired targets, outcomes and indicators relevant to their context across the full breadth of opportunity. In addition, UNESCO and *UNFCCC* (2016)[63] recommend engaging diverse stakeholders when developing a national strategy. However, *ACE* involves stakeholders across the whole of the economy, making meaningful consultation and engagement an intensive and complex task.

[60] UNESCO, *Action for Climate Empowerment: Guidelines* (Footnote 31).

[61] UNESCO, *Integrating Action for Climate Empowerment* (Footnote 33).

[62] Martha C. Monroe et al., 'Identifying Effective Climate Change Education Strategies: A Systematic Review Of Research' (2017) 25(6) *Environmental Education Research* 791. https://doi.org/10.1080/13504622.2017.1360842.

[63] UNESCO, *Action for Climate Empowerment: Guidelines* (Footnote 31).

Additionally, notwithstanding *ACE*'s foundational principles, detractors contend that there is a deficiency in a dependable and trustworthy procedure for establishing targets and monitoring progress within *ACE*.[64] Moreover, they argue that national governments may not consistently have access to the essential information or support required for the successful execution of *ACE* initiatives. Related to this, not all countries currently want to, or have the capacity to, develop a national plan, which is often a reflection of the finance and prioritisation of *ACE*. Finance is a particular concern for developing nations with regard to *ACE* and remains a cross-cutting issue for many other areas of climate mitigation, adaptation and resilience.

6.3 Youth and *Action for Climate Empowerment* for a Just Transition

There are many elements of youth and *ACE* that could be discussed in this chapter in relation to a just transition; however, this section focusses on access to quality climate change education, youth perspectives on *ACE* and meaningful youth participation in decision-making.

6.3.1 Quality Climate Change Education as a Core Component of ACE

There are over 1.8 billion young people between the ages of 10 and 24 years today.[65] Many of them are grappling with increased levels of uncertainty about the future and want the opportunity to contribute to a just transition by shaping solutions. As stated by United Nations Secretary-General Antonio Guterres:

[64] Climate Outreach, 'Principles and Process for Tracking Progress in UNFCCC's Action for Climate Empowerment (ACE)'. https://climateoutreach.org/principles-process-tracking-progress-unfccc-ace/.

[65] UN News Global Perspective Human Stories, 'World's 1.8 Billion Youth Must Have a Say in the Future of the Planet'. https://news.un.org/en/story/2019/09/1046882.

> My generation has largely failed until now to preserve both justice in the world and to preserve the planet. It is your generation that must make us be accountable to make sure that we don't betray the future of humankind.[66]

Access to comprehensive climate change education in both formal and informal learning settings serves as the foundational pillar for impactful climate action. Education is widely recognised as the most powerful instrument for enabling individuals and communities to take meaningful steps towards addressing climate challenges.[67] Educational initiatives focussed on enhancing knowledge and skills cultivate responsible citizens and capable leaders who can spearhead holistic shifts in our economic systems.[68] However, mainstreaming compulsory climate education remains a challenge. While countries such as Sweden, Italy, New Zealand and Mexico have introduced mandatory climate and environmental education,[69] access remains an issue in many countries. In general, recommendations related to *ACE* and the *UNFCCC* have not yet trickled down to mainstream curriculum transformations, with national frameworks required to drive real action.

Defining quality climate change education is also difficult. There is a need to create global indicators for what quality climate change education encompasses and how it should be measured in NDCs. This includes creating a greater understanding among providers that quality climate education extends beyond cognitive learning and scientific knowledge to cultivate socio-emotional and behavioural proficiencies necessary for intervening in intricate social systems, thus propelling social transformation. According to the National Oceanic and Atmospheric Administration's definition,[70] 'You are climate literate if you understand the influence of

[66] Ibid.

[67] United Nations Climate Action, 'Cities and Local Action: Education is Key to Addressing Climate Change'. https://www.un.org/en/climatechange/climate-solutions/education-key-addressing-climate-change.

[68] Ibid.

[69] UN News Global Perspective Human Stories, 'COP25: Italy and Mexico Pave the Way on Climate Education'. https://news.un.org/en/story/2019/12/1053261.

[70] US Global Change Research Program, *Climate Literacy: The Essential Principles of Climate Science* (National Oceanic and Atmospheric Administration and the American Association for the Advancement of Science, 2009).

climate on yourself and society—and your influence on climate'. Further, a global lens is required with consideration of inequality. Cross-cultural learning is a necessity, as future change-agents will need to work from the perspectives of their own cultural landscapes, while honouring those brought to the table by diverse multidisciplinary international peers from both sides of socioeconomic and political divides. At the 2023 Mock COP, youth from 89 countries developed the *Youth Statement on Quality Climate Education*, arguing that education can equip learners to 'build a just, equitable and sustainable future' and that the current education system is 'unable to prepare us to face the climate crisis or understand its solutions'.[71] They call for a comprehensive, solutions-focussed, interdisciplinary, intersectional and inclusive curriculum that can empower learners to decarbonise societies.

Quality climate change education (and the intersecting area of Education for Sustainable Development)[72] is a lifelong process. It takes place in schools and universities but also through many other avenues including public engagement and literacy programs and professional development and training. We need to renew our focus on addressing SDG 4 with regard to lifelong learning[73] to ensure the ongoing resilience and adaptive capacity of communities. Learners must be regularly equipped and updated with the latest knowledge and solutions. Initiatives must be specific to the local context, featuring participation in local policy generation, co-production of research projects with stakeholders and immersion in industry challenges using the SDG framework in the language of international governments, industries and local communities. This adult education requires collaboration across sectors

[71] Students Organising for Sustainability UK, *Youth Launch Their Unified Statement on Climate Education,* 16 August 2023, (SOS UK, 2023). https://www.sos-uk.org/post/youth-launch-their-unified-statement-on-climate-education.

[72] UNESCO, *Changing Minds, Not the Climate: The Role of Education* (UNESCO, 2017). https://unesdoc.unesco.org/ark:/48223/pf0000266203?posInSet=4&queryId=e23e02bb-27a2-4f38-a696-231658a9f315.

[73] UNESCO, 'Sustainable Development Goal 4 and its Targets'. https://en.unesco.org/education2030-sdg4/targets#:~:text=By%202030%2C%20eliminate%20gender%20disparities,and%20children%20in%20vulnerable%20situations.

and cultures in alignment with SDG 17 (Partnerships for the Goals).[74] To ensure that opportunities are relevant and responsive to learners' contexts, educators must partner with the public and private sectors as well as local communities to co-design and tailor initiatives for a spectrum of learners across different life and career stages.

6.3.2 *Youth Perspectives on Action for Climate Empowerment*

In 2009 the *UNFCCC* secretariat granted constituency status to youth and youth organisations known as YOUNGO. YOUNGO operates as a worldwide network of young individuals and activists up to the age of 35 years, along with youth-focussed non-governmental organisations.[75] Their primary objectives include contributing to international climate change policies within intergovernmental frameworks and empowering young people to actively voice their concerns in *UNFCCC* processes. YOUNGO representatives assume various responsibilities including making official statements, offering technical and policy insights during negotiations and engaging with decision-makers to advocate for increased youth involvement in climate change initiatives at local, sub-national, national, regional and international levels.[76]

YOUNGO is an established supporter and advocate of *ACE*. They have advocated for it in various ways, recognising its value to the goals of youth organisations, including climate justice. For example, the *Global Youth Statement*[77] led by YOUNGO[78] calls on parties to:

> acknowledge the significance of ACE as a key pillar of climate action through adopting a robust ACE Action Plan with universal metrics and attributions to each element. Additionally, Parties must incorporate the appointment of national ACE youth coordinators or focal points to work in partnership with National ACE focal points and address the ACE finance gap by convening funders under one platform in the ACE action plan.

[74] United Nations Department of Economic and Social Affairs, Sustainable Development, 'Goal 17'. https://sdgs.un.org/goals/goal17.

[75] YOUNGO, 'The Official Children and Youth Constituency of the United Nations Framework Convention on Climate Change'. https://youngoclimate.org/.

[76] Ibid.

[77] Ibid.

[78] YOUNGO, 'The Conference of the Youth'. https://youngoclimate.org/about/.

This speaks to the fact that while youth support *ACE*, they recognise that it is not being implemented to the same degree across all countries and cannot be fully realised without youth inclusion. For example, some countries have implemented national strategies while many have not. Insufficient finance for *ACE* implementation is recognised by YOUNGO as a concern for young people to access capacity building and participation opportunities as well as governments to reach their climate action goals. Without a two-way exchange between young people and senior decision-makers, young people do not feel decisions today have their best interests in mind.

YOUNGO's main source of advocacy within ACE has been utilising its foundational principle of participation as a tool, leading to greater youth participation, including through the activities of the YOUNGO ACE Working Group. As an illustration, some notable achievements of YOUNGO regarding the *Glasgow Work Programme* encompass organising the initial in-session workshop at SB 56, which emphasised youth involvement; garnering backing for an *ACE* implementation plan; receiving support from parties, the COP Presidency and the secretariat for hosting an annual *ACE* Youth Forum and ensuring the participation of young individuals within delegations, either as negotiators or National Youth *ACE* Focal Points.[79]

6.3.3 *Meaningful Youth Engagement in Decision-Making*

A clear finding of the 2022 UNICEF Climate Justice Roundtable[80] was that young people want a seat at the table to achieve climate justice. The roundtable report calls on decision-makers to support 'their meaningful participation'.[81] Taking a climate justice approach to *ACE* includes integrating youth perspectives and youth rights to reduce their vulnerability to the climate crisis. This aligns with the goal of Article 6 in ensuring governments and organisations around the world take youth perspectives into account when making policy decisions. UNICEF recognises the importance of enabling and widening participation in decision-making

[79] COY17 and YOUNGO, *COP27 Global Youth Statement* (Footnote 24).

[80] UNICEF, *Recap and Reflections* (Footnote 18).

[81] Ibid.

processes by financing the capacity and skills building of youth as well as building partnerships to help climate justice action to flourish.

Despite some progress, obstacles still exist when it comes to effectively incorporating the perspective of young people into the most influential levels of decision-making. For example, at the international level, given the COP's pivotal role in steering global diplomacy and assembling representatives from 198 countries, it is essential that young people, who are stewards of the future, have their voices heard.[82] Engaging with the COP also offers a distinct avenue for cultivating future leaders in areas such as global governance through fostering cross-cultural connections and collaborations. However, youth have historically been asked for advice outside formal settings and report being placed in positions that might appear symbolic or as mere gestures of inclusivity during COP events. The term 'youthwashing' has emerged to reflect the use of youth to make organisations look good as opposed to being directly involved and heard in official negotiations.[83]

There is a need to develop enhanced intergovernmental, as well as national, sub-national and local mechanisms that allow youth to substantially contribute to formal policies and processes. Co-contact point for YOUNGO's ACE Working Group, Hailey Campbell, said in a 2022 interview with *CNBC Make it*, that[84]:

> When we talk about representation, we don't just want it at international negotiations and we don't want to only be consulted. We want it at all levels of government and we want to be partners because action happens on the ground.

[82] Gillian Bowser et al., 'The Fierce Urgency of Now: Integrating the Youth Voice at COP' (2023) *Frontiers in Ecology and the Environment*. https://doi.org/10.1002/fee.2627.

[83] The Climate Reality Project, 'Blog: You've Heard of Greenwashing but What is Youthwashing?' (Blog Post, 5 April 2023). https://www.climaterealityproject.org/blog/youve-heard-greenwashing-what-youthwashing.

[84] Sophie Kiderlin, 'Young People Just Got a Louder Voice on Climate Change—and Could Soon be Shaping Policy' *CNBC Make it*, 23 November 2022. https://www.cnbc.com/2022/11/23/young-people-just-became-official-climate-policy-stakeholders-at-cop27.html.

She argues that including more youth in decision-making 'creates more ambitious and just outcomes',[85] pointing to the fact that youth involvement might advance rapid action. In many cases, there may be good will from decision-makers, who may struggle with the complexity of retrofitting existing policy and consultation structures to incorporate a broader range of youth input.

Youth will also require specialised training to use their seat at the table effectively. For example, the complexity of its policy language, sheer size, chaotic structure and intricate intergovernmental procedures can make the COP appear daunting and difficult to engage with. Without training and mentorship, COP is challenging to navigate for students and seasoned professionals alike. Given this, a plethora of climate leadership programmes and online 'beginners guides' have emerged in the last five years to equip stakeholders to navigate global climate diplomacy. Nevertheless, access to these programmes can be difficult for some youth, exacerbated by the fact that many young people cannot financially afford to travel to other countries for COPs. Furthermore, passes to enter COPs are in short supply. Similar access issues apply at local, national and sub-national levels.

Nonetheless, there has been notable headway in incorporating the perspectives of young individuals and promoting climate justice. The year 2022 marked a significant step forward in this regard during COP27.[86] This milestone event featured the inaugural participation of an official youth representative, Omnia El Omrani, who fervently advocated for the inclusion of young voices. Additionally, it witnessed the launch of a programme aimed at nurturing young climate activists from the Global South to become climate negotiators, and the establishment of the first-ever youth climate forum.

Further, 2022 marked a pivotal moment as youth attained official status as stakeholders in climate policy in the *ACE Action Plan*. This development means that the voices and viewpoints of young people will carry considerably more weight in the formulation and execution of climate policies. Hailey Campbell elucidates that this formal recognition

[85] Ibid.

[86] Ulrika Modéer et al., 'Celebrating COP27 Wins for Young People' *United Nations Development Programme*, (Blog Post, 25 January 2023). https://www.undp.org/blog/celebrating-cop27-wins-young-people.

as stakeholders in the plan 'provides young people with the international support necessary to insist on our formal involvement in climate decision-making and execution'.[54]

6.4 Conclusion

ACE, a crucial component of both the *UNFCCC* and the *Paris Agreement*, serves as a cross-cutting framework spanning various climate policies. It offers a vital and multifaceted avenue for empowering youth in the pursuit of climate justice. However, its effectiveness hinges on the ability of participating parties to appropriately prioritise, finance, activate and monitor *ACE* through their national action plans, with the support of international cooperation. The coming times will reveal whether *ACE* can secure the necessary funding and coordination within a reasonable timeframe.

Youth play a pivotal role as stakeholders in *ACE*, vigorously championing this cause alongside other critical climate issues, particularly climate justice and a just transition to a low-carbon economy. While they are steadily gaining ground in their efforts to advocate for climate justice and ensure their voices are heard in decision-making processes, it remains imperative that we extend our support to ensure they have a meaningful presence in these discussions.

Mandatory high-quality climate change education emerges as a critical tool to bolster the youth movement. Ideally, we should expand access to comprehensive educational programmes that not only enrich cognitive understanding but also nurture socio-emotional and behavioural skills. These programmes should be designed to cater to lifelong learning needs, involving collaboration with communities, industries and government bodies. Further, exposing students to diverse cultural contexts should be an integral part of this education. This approach will empower them to become competent, just and collaborative leaders before they embark on their career paths and life journeys.

As *ACE* transitions into the implementation phase, it is crucial to ensure that youth receive quality education and training. They should also be provided with expanded opportunities to participate in decision-making processes at various levels—be they local, national, sub-national or intergovernmental.

CHAPTER 7

Conclusion

Abstract Chapter 1 introduced the challenge of a just transition to decarbonisation. Chapter 2 argued that creating a loss and damage fund with the purpose of providing support and assistance for vulnerable countries demonstrates a first move towards recognising climate justice. Chapter 3 on the transport theme posited that more taxation support in Australia is necessary to achieve zero-emissions transport, with a focus on zero-emissions cars in business fleets. Chapter 4 on the nature theme reasoned that it is vital for a just transition to decarbonisation that First Nations Peoples' water knowledge be recognised in Australia's water law and policy. Chapter 5 on the theme of nature, questioned whether free trade agreements (FTAs) should or could be utilised in redressing the harm inflicted by climate change upon the natural environment. Chapter 6 centred on empowering youth and argued that action for climate empowerment (ACE)—a crucial component of both the United Nations Framework Convention on Climate Change (UNFCCC) and the Paris Agreement—serves as a cross-cutting framework spanning various climate policies.

Keywords A just transition · Loss and damage · Transport · Nature · Youth · Energy justice · Climate justice

D. Kraal et al., *A Just Transition to Decarbonisation*, Just Transitions,
https://doi.org/10.1007/978-3-031-49137-5_7

This book's remit has been to advance the need for the global transition to decarbonisation in a way that is 'just'. The challenge of this transition is the challenge of our lifetime. This book has advocated the urgency for a just transition to 2030 and beyond. It is observed that the ambition of the majority of countries across the world is to develop a low-carbon economy. This is evidenced by the vast majority of countries having signed the 2015 Paris COP21 agreement.

To transition away from a carbon-based economy the process must be just and equitable on a local to global scale. Recalled is the experience of the North American trade union movement's aim to secure workers' rights and livelihoods as economies shift away from coal and other carbon-producing energy to sustainable resources. The union made the clarion call that—a just transition to decarbonisation—means: leave no one behind. States globally have to rethink the implications on the operation of their economic systems. It is crucial that all key stakeholders be involved, including the least developed states, Indigenous populations and non-state actors. Thus, the vulnerable, the young and the wider community within any state irrespective of its level of development must be given a voice and heard. This imperative has resonated throughout this book.

At COP27 Egypt in 2022 the concept of structuring discussions continued around thematic ideas; such as finance, science, youth and future generations; decarbonization, adaptation and agriculture; gender, water, action for climate empowerment and civil society; loss and damage; energy and biodiversity solutions. In 2023 at the COP28 meeting in Dubai, the parties adopted and extended these many structural themes. Thus, this book's chapters have been based around four enduring COP themes in particular: loss and damage, transport, nature and youth.

These four themes will scaffold discussions about a just transition beyond COP28 in Dubai. It is expected that their relevance will remain, and that the issues addressed in this book will serve as a starting point for future discussions.

This book demonstrates that justice scholarship can vary, whether it be climate justice, energy justice or environmental justice. Some chapters have further observed justice as distributional, procedural, restorative, while needing both recognition and cosmopolitanism. Justice is an increasingly important issue across all areas of climate action. From taxation to biodiversity, justice issues are inherent. It has been argued in this book that various strategies have to be employed at sub-national, national and international levels to achieve a just transition to decarbonisation.

Moreover, battles have to be fought on all fronts, across the whole of the economy, to address this complex and multifaceted issue.

Chapter 1 introduced the challenge of a just transition to decarbonisation. Chapter 2 looked at the history and development of the issue of loss and damage at intergovernmental negotiations. It was observed that the decision to establish a dedicated fund in 2022 at COP27 Egypt marks a historic breakthrough in climate change history as it follows years of widespread pressure from developing nations to compensate them for the harm caused by wealthy nations. The establishment of a fund is seen as a significant shift, after many years of reluctance by big international influencers, including the European Union and the US. However, the chapter cautioned that even though the COP27 meeting called the fund a 'step towards justice', it is a very small step. It remains unclear at this juncture how the fund will be effectively implemented and administered, particularly in the absence of an agreed definition of loss and damage. Caution needs to be exercised vis-à-vis the effectiveness of the fund, given the current exclusion of all questions concerning liability and compensation within the legal instruments created for loss and damage.

Chapter 2 then argued that creating a loss and damage fund with the purpose of providing support and assistance for vulnerable countries demonstrates a first move by the global community towards recognising climate justice, but it cannot be achieved unless those responsible for climate change are held morally and legally accountable. However, no meaningful agreement has been made on funding as developed states maintain that they are not bound to compensate and will not suffer legal liability. In effect, no real shift has occurred in terms of developed states taking legal responsibility for their significant share of carbon emissions. Therefore, the funds creation may potentially be viewed as tokenistic, demonstrating that there is still work to be done towards fully integrating and operationalising an effective climate justice template. While a loss and damage fund will not necessarily change the trajectory of the climate crisis, its successful implementation is seen to be beneficial for affected vulnerable communities. This should not be overlooked in the just transition to decarbonisation debate.

A global reduction in carbon emissions is necessary to minimise further loss and damage. At the practical level, one way that economies can change behaviour in regard to excessive emissions is through the taxation toolkit of concessions and subsidies. Chapter 3 on the transport theme posited that more taxation support in Australia is necessary to achieve zero-emissions transport, with a focus on zero-emissions cars in business fleets. There are clear decarbonisation benefits of increasing zero-emissions transport. In 2021 the Australian government identified the light vehicle transport segment, and business fleets in particular, for national CO_2 reductions. It successfully passed some relevant legislation. It was discussed how more Australian taxation changes are needed to further facilitate the acquisition of affordable zero-emission cars for business and associated charging infrastructure. The chapter provided European tax comparisons for benchmarking purposes. For instance, Norway's tax policies have had the most positive impact on electric vehicle (EV) uptake in Europe. Its VAT concessions for cars and purchase subsidies were key. Germany offers some of the most generous EV incentives in Europe, with a goal of having 10 million EVs and 1 million charging stations by 2030. Interestingly, The Netherlands has the highest number of EV charging points in the EU. The UK's national laws mirror that of EU regulations for car emission performance standards; along with its innovative capital allowance and congestion tax concessions for BEVs.

The Australian government's National Electric Vehicle Strategy 2023 is designed to increase the uptake of EVs and introduce emissions standards. This will be a test of whether new policy will result in further legislation. This can be beneficial at a practical level, as business EV fleet cars can drive a just transition to decarbonisation to the benefit of the wider community via the second-hand car market.

Two chapters in this book took up the COP theme of nature. Both chapters looked at issues of relevance to Indigenous communities, including water, biological diversity and the interrelated issue of biocultural heritage. Notably, the issues raised in relation to Australia can provide lessons for other jurisdictions as well.

Chapter 4 on the nature theme argued that it is vital for a just transition to decarbonisation that First Nations Peoples' water knowledge be recognised in Australia's water law and policy. Globally, it argued that Indigenous communities' water access be recognised and protected to ensure that these communities are able to maintain their spiritual and cultural connection to their lands and waters. Furthermore, Indigenous communities should also have the opportunity to actively contribute to shaping sustainable water management. It stated that some steps in this direction have been made recently in Australia, particularly at a policy level. However, they are still small and fragile steps as a shift in policy has not yet been translated into meaningful legislative reform, thus making

First Nations Peoples' water knowledge vulnerable to changing political priorities. It is hoped, however, that support for change will improve over time.

Chapter 5 also on the theme of nature, questioned whether free trade agreements (FTAs) should or could be utilised in redressing the harm inflicted by climate change upon the natural environment. In particular biodiversity—and its concomitant effect on the intertwined cultural diversity, heritage and vulnerable Indigenous peoples, and thereby have the potential to contribute to a just transition to a low emission and resilient economy. As climate change and biodiversity crises worsen there is the urgency that they be tackled on various fronts, including trade due to its vast prominence in the global economy. Given that FTAs are being increasingly negotiated throughout the world, the question of their utility vis-à-vis the protection of nature is becoming ever more significant. By taking a biocultural heritage approach, it is argued that in transitioning away from a fossil fuel economy, FTAs have the potential to play an important role in the context of international trade, climate change, biodiversity and heritage, and that this transition cannot be seen as just without an adequate place being given to Indigenous peoples.

The final theme on youth was covered by Chapter 6 that centred on the importance of empowering youth in a just transition to decarbonisation. It stated that if we are to move away from a carbon-based society to a sustainable future, then it must be seen as just across generations. Those such as youth will be most impacted by climate change and they must be given a voice in a meaningful way. The chapter argued that action for climate empowerment (ACE)—a crucial component of both the *United Nations Framework Convention on Climate Change* (UNFCCC) and the *Paris Agreement*—serves as a cross-cutting framework spanning various climate policies. ACE offers a vital and multifaceted avenue for empowering youth (and the public) to accelerate climate justice. However, its effectiveness depends on the ability of member parties, among other things, to get finance and monitor ACE through their National Action Plans.

Finally, the chapter observed that youth play a pivotal role as stakeholders in ACE, vigorously championing it alongside complementary issues, such as climate justice. Youth's involvement is essential in decision-making through relevant avenues, and in discussions at regional, national and international levels. The chapter covered ways to enhance youth's involvement in a just transition, such as through mandatory high-quality climate change education. The power of education, together with climate change literacy, is a critical tool to bolster the youth movement. It is noted that youth should have access to comprehensive educational programmes that not only enrich cognitive understanding but also nurture socio-emotional and behavioural skills. This will develop multicultural, competent, collaborative, global leaders equipped to take individual and collective action.

The effectiveness of a just transition will ultimately depend on the engagement of today's global community. Just as every single drop in the ocean counts, each of us, too, can make a valuable contribution by rethinking our lifestyles and seeking to reduce our own carbon footprint. We can do this by adopting sustainable practices in our daily lives, and by educating ourselves and others about the importance of preserving life support systems and ensuring human and environmental wellbeing. This book has emphasised the paramountcy of a just transition to decarbonisation in a way that is not only sustainable but also equitable so that present as well as future generations are not disadvantaged. This is our legal, moral and humane duty which is owed to the generations that follow us.

Index

D. Kraal et al., *A Just Transition to Decarbonisation*, Just Transitions,
https://doi.org/10.1007/978-3-031-49137-5

SPRINGER NATURE

GPSR Compliance

The European Union's (EU) General Product Safety Regulation (GPSR) is a set of rules that requires consumer products to be safe and our obligations to ensure this.

If you have any concerns about our products, you can contact us on ProductSafety@springernature.com

In case Publisher is established outside the EU, the EU authorized representative is:

Springer Nature Customer Service Center GmbH
Europaplatz 3
69115 Heidelberg, Germany